Mathematical Analysis

M. D. Hatton M.Sc., F.I.M.A.

Senior Lecturer in Mathematics
University of Surrey

HODDER AND STOUGHTON

LONDON SYDNEY AUCKLAND TORONTO

ISBN 0 340 20824 4 (Boards)
0 340 20825 2 (Unibook)

First printed 1977

Printed in Great Britain for
Hodder and Stoughton Educational,
a division of Hodder & Stoughton Ltd.,
Mill Road, Dunton Green, Sevenoaks, Kent,
by J.W. Arrowsmith Ltd., Bristol.

Preface

This text has been designed as an introduction to basic classical analysis. Starting with the idea of irrational numbers it develops the concepts of limit, continuity (including uniformity of behaviour) and covergence, Riemann integration and an introduction to functions of several variables. There is no mention of more abstract spaces as most students need to be acquainted with these ideas on the real line before they can appreciate generalizations to metric and topological spaces.

I have assumed that the reader will have a knowledge of elementary techniques of calculus. The book is aimed primarily at mathematics students, but it is hoped that it will also prove of value to students of engineering and science.

The exercises are an essential part of any mathematical text and the reader is urged to attempt as many as possible; hints are given for the more difficult ones.

I would like to thank the Senate of the University of London for permission to use questions from their examination papers. I am particularly indebted to my colleague Dr. R.S. Taylor for his helpful comments on the typescript and for his help in checking the exercises.

M.D. Hatton.

Contents

CHAPTER 1

Preliminaries

1.1 Introduction

Mathematical analysis is that branch of pure mathematics concerned with limits; this includes such topics as continuity of functions, differentiation, integration and convergence of infinite series.

A fundamental concept underlying all these topics is that of number. We make no attempt to define this abstract idea,* and take as our starting point the *rational numbers*. That is numbers of the form p/q, where p and q are integers and $q \neq 0$. We assume that the reader is familiar with the rules for the arithmetical manipulation of these numbers. The sum, difference and product of two rationals are rational; the quotient of two rationals is rational, except that division by zero is not permitted. Both addition and multiplication of rationals are commutative and satisfy the associative and distributive laws. These properties can be summarised by saying that the rational numbers form a *field*. Details of the development of the rationals from the natural numbers (the positive integers) may be found in many algebra books, such as C.C. MacDuffee, *An Introduction to Abstract Algebra,* Wiley, 1948.

The rationals are *ordered* by the relation $<$, which has the property that, for any two different rationals p, q $(p \neq q)$, either $p < q$ or $q < p$. If b and d are positive integers we define $a/b < c/d$ to mean $ad < bc$. Also, the rationals are *dense* in any interval, that is there is no rational which comes immediately after any given rational. In other words, between any two rationals there is another $(p < \frac{1}{2}(p+q) < q)$ and so infinitely many others.

* See for example B. Russell, *Introduction to Mathematical Philosophy*, Allen and Unwin, 1920.

1.2 Irrational numbers

However, there are 'gaps' in the system of rationals. This is easily demonstrated by showing that $\sqrt{2}$ is not rational. Such a number is called *irrational*.

Suppose $\sqrt{2} = p/q$, where p, q are integers with no common factor. Then $p^2 = 2q^2$, and so p^2 is even; hence p is even, say $p = 2k$. Substituting for p we have $q^2 = 2k^2$. Hence q is even also; but this contradicts the assumption that p and q have no common factor. Therefore $\sqrt{2}$ is not rational.

The existence of other irrationals, such as $\sqrt{3}, \sqrt[3]{2}, \sqrt[4]{5}$, can be established by similar arguments.

Using the usual arithmetical method for the extraction of square roots we can find a succession of rational numbers

$$1{\cdot}4,\ 1{\cdot}41,\ 1{\cdot}414,\ 1{\cdot}4142,\ 1{\cdot}41421,\ 1{\cdot}414213, \ldots$$

whose squares are less than 2* by

$$4 \times 10^{-2},\ 1{\cdot}19 \times 10^{-2},\ 6{\cdot}04 \times 10^{-4},\ 3{\cdot}836 \times 10^{-5},$$
$$1{\cdot}008 \times 10^{-5},\ 1{\cdot}591 \times 10^{-6}, \ldots .$$

Since there is no rational whose square is 2, the positive rationals fall into two classes, those whose squares are greater than 2 and those whose squares are less than 2.

Consider L, the class of positive rationals whose squares are less than 2. If l is any member of L, then we can find another rational in L, greater than l; that is, L has no greatest member. For if l is any member of L we can choose a rational h, $0 < h < 1$, such that $p = l + h$ is also a member of L. Thus,

$$p^2 = l^2 + h(2l + h) < l^2 + h(2l + 1) < l^2 + 2 - l^2 = 2$$

provided $$h < (2 - l^2)/(2l + 1).$$

(See also Exercises 1(a) No. 4.)

Similarly, R, the class of positive rationals whose squares are greater than 2 has no least member. (See Exercises 1(a) No. 3.)

Clearly any member of L is less than every member of R. Also, it is possible to find a member of L and a member of R which differ by as little as we please. Let δ be any small positive number and l, r members of L and R respectively, such that $2 - l^2 < \delta$ and $r^2 - 2 < \delta$. Then

* We could, of course, write down a succession of rationals whose squares exceed 2; thus

$$1{\cdot}5,\ 1{\cdot}42,\ 1{\cdot}415,\ 1{\cdot}4143,\ 1{\cdot}41422,\ 1{\cdot}414214, \ldots .$$

$$r^2 - l^2 = (r^2 - 2) + (2 - l^2) < 2\delta,$$

and, since $l > 1$ and $r > 1$, we must have

$$r - l = \frac{r^2 - l^2}{r + l} < \delta.$$

Now let α be the set of rational numbers obtained by adding the negative rationals to elements of L. Then, if a is any rational in α, it is clear that any rational less than a must also be in α. Also, as proved above, α has no greatest member. We say that the set of rationals α *defines* the irrational number $\sqrt{2}$, and that α is a *section* of the rational numbers. This process can be used to define any irrational number.

1.3 Dedekind Sections

Dedekind's definition of irrational numbers is based on sections of the rationals, that is a set α of rationals such that

(a) α contains at least one rational but not every rational;
(b) if r is a rational in α, then every rational less than r is also in α;
(c) α contains no greatest member.

A section of the rationals defined in this way divides the set of all rationals into two classes; the members of α, which may be called the *lower class* of rationals, and those rationals not in α called the *upper class.* The upper class may or may not have a smallest member; if its possesses a smallest member the section is called a *rational section*.

If r is any rational, then the set α defined by taking all the rationals x such that $x < r$ is a rational section. Clearly (a) and (b) of the definition of a section are satisfied. Also, if x is in α, so also is $\frac{1}{2}(x + r)$, since $x < \frac{1}{2}(x + r) < r$. Further α is a rational section; r is not in α and so must be the least member of the upper class corresponding to α.

Each section of the rationals is said to define a *real number*. The rational sections represent the rational numbers, so that the rationals may be identified with a subset of the real numbers; all other sections are called irrational numbers. We must now show that the set of all sections, that is, the set of real numbers, satisfies the same laws of addition, multiplication, etc. as do the rationals. These laws are listed in Section 1.5.

First we define *equality* of sections; two sections α, β are said to be equal, $\alpha = \beta$, if they are identical sets of rationals. That is, any member a of α is also in β and any b in β is also in α.

If $\alpha \neq \beta$, then either $\alpha < \beta$ or $\alpha > \beta$, the former being true when there is a rational b in β which is not in α.

The sum of two sections α, β is the set γ of all rationals c such that $c = a + b$, where a is in α and b in β. We write $\gamma = \alpha + \beta$. It must, of course, be verified that γ is indeed a section, by showing that (a), (b) and (c) of the above definition of a section are satisfied. Since α, β are sections, there is a rational p not in α and a rational q not in β. Then for all a in α, b in β, $a + b < p + q$, and so $p + q$ is not in γ; hence (a) is satisfied.

To show (b) is satisfied let a be in α, b in β and $c = a + b$ in γ. For any rational $p < c$, we can find a rational q such that $p = q + b$. Hence $q < a$ and q is in α, and so p is in γ.

Let $c = a + b$ for some a in α, b in β. Since α is a section there is a rational $p > a$ in α. Then $p + b$ is in γ and $p + b > c$. That is γ has no greatest member and so γ is a section.

The commutative and associative laws of addition follow directly from the corresponding properties of the rationals. For $\alpha + \beta$ is the set of rationals $a + b$, and $\beta + \alpha$ the set $b + a$ where a is in α and b is in β.

Multiplication and division of sections may be considered in analogous ways; for details the reader is referred to G.L. Isaacs, *Real Numbers*, McGraw-Hill, 1968, or to E.G.H. Landau, *Foundations of Analysis*, Chelsea, 1951.

1.4 Dedekind's theorem

Theorem (Dedekind) If the real numbers are divided into two classes L and R such that

(a) every real number is either in L or in R;
(b) no real number is in both L and R;
(c) neither L nor R is empty;
(d) every member of L is less than every member of R,

then there is one and only one real number ξ, such that every real number less than ξ belongs to L and every real number greater than ξ belongs to R.

The number ξ may belong to either L or R. If it belongs to L it is the largest member of L; if it belongs to R it is the smallest member of R.

Let ξ be the set of all rationals a such that a is in α for some α in L. Then

(a) ξ is not empty, for L is not empty; ξ does not contain all the rationals. Let β be in R and p a rational not in β, then p cannot be in any α in L, for $\alpha < \beta$ and $p > a$ for every rational a in α.

(b) Let a be in ξ, then a is in α for some α in L. Then if $p < a$, p is in α and so p is in ξ.
(c) Let a be in ξ, then a is in α for some α in L. Since α is a section, α has no greatest member. Hence ξ has no greatest member.

Therefore ξ is a section of the rationals, that is, ξ is a real number. Also it it is unique; for if there were two such numbers ξ_1, ξ_2 and $\xi_1 < \xi_2$, we could find a number ξ_3, such that $\xi_1 < \xi_3 < \xi_2$. Now $\xi_1 < \xi_3$ implies ξ_3 is in R, whereas $\xi_3 < \xi_2$ implies ξ_3 is in L, which contradicts (b).

This theorem shows the *completeness* of the real numbers, that is, there are no 'gaps' such as the rationals possess. If the process of defining sections (Section 1.3) is applied to the real numbers, then for every such section, R contains a smallest member. Thus every section could be identified with this smallest member of R (corresponding to rational sections, Section 1.3) and so no new number would be obtained.

1.5 Real numbers

For convenience of reference we list in this section the properties of the real-number system **R**. Let a, b, c represent any real numbers:

(1) $a + b = b + a$.
(2) $(a + b) + c = a + (b + c)$.
(3) There is a zero 0 such that $a + 0 = 0 + a = a$.
(4) For each a there is a real number $-a$ such that $a + (-a) = (-a) + a = 0$.
(5) $ab = ba$.
(6) $(ab)c = a(bc)$.
(7) There is a unit element 1 such that $1a = a1 = a$.
(8) For each $a \neq 0$ there is a real number a^{-1} such that $aa^{-1} = a^{-1}a = 1$.
(9) $(a + b)c = ac + bc$.
(10) For each pair of real numbers a, b only one of $a = b$, $a < b$, $a > b$ holds.
(11) If $a < b$ and $b < c$, then $a < c$.
(12) If $a < b$, then $a + c < b + c$.
(13) If $a < b$ and $c > 0$, then $ac < bc$.

These properties show that the real numbers form an *ordered field*.

EXERCISES 1(a)

1. Prove that there is no rational number r such that

$$(a)\ r^2 = 3 \quad (b)\ r^2 = 5 \quad (c)\ r^3 = 2.$$

2. Find the differences between 2 and the squares of the numbers given in the footnote on p. 10.

3. Prove that the set of positive rationals whose squares are greater than 2 has no least member.

4. By considering the rationals p and q satisfying the relation $q = p(p^2 + 6)/(3p^2 + 2)$, show that L (defined in Section 1.2) has no greatest member and R no least member.

5. Prove that if a is a rational member of a section α and b is a rational which is not in α, then $a < b$.

6. Prove that if α, β, γ are sections of rationals such that $\alpha < \beta$ and $\beta < \gamma$, then $\alpha < \gamma$.

7. Prove that if α is a section of rationals and 0 the section corresponding to zero, then $\alpha + 0 = \alpha$.

8. Prove that if α, β are sections and $\alpha < \beta$, then there is a rational section γ, such that $\alpha < \gamma < \beta$.

1.6 Mathematical Induction

The principle of *mathematical induction* may be stated:

Let $P(n)$ stand for the proposition to be proved. If (a) the truth of $P(n)$ implies the truth of $P(n+1)$, and (b) $P(n)$ is true for a particular value k of n (usually $k = 1$ or 2), then $P(n)$ is true for every positive integer $n \geqslant k$.

Example 1 Prove $1^2 + 2^2 + 3^2 + \cdots + n^2 = \frac{1}{6}n(n+1)(2n+1)$.

Let $P(n)$ stand for the formula to be proved. Suppose $P(n)$ is true, then

$$\begin{aligned} 1^2 + 2^2 + 3^2 + \cdots + n^2 + (n+1)^2 &= \tfrac{1}{6}n(n+1)(2n+1) + (n+1)^2 \\ &= \tfrac{1}{6}(n+1)[n(2n+1) + 6(n+1)] \\ &= \tfrac{1}{6}(n+1)(n+2)(2n+3). \end{aligned}$$

Thus, if $P(n)$ is true so also is $P(n+1)$. Clearly the formula is true for $n = 1$; therefore the result is true for any positive integer n.

It is important to remember that a proof by induction is in two parts; it is necessary to establish the hereditary property *and also* to show that $P(n)$ is true for a particular value of n, such as $n = 1$.

Example 2 Is $P(n) = 3^{2n} - 8n + 1$ divisible by 64?

Consider

$$\begin{aligned} P(n+1) - P(n) &= 8(9^n - 1) \\ &= 64(9^{n-1} + 9^{n-2} + \cdots + 9 + 1). \end{aligned}$$

Thus, if $P(n)$ is divisible by 64 so also is $P(n+1)$. But $P(n)$ is *not* divisible by 64; for we cannot find a starting value for the 'truth' of $P(n)$.

See also Exercises 1(b) No. 10 for another example of fallacious use of induction.

EXERCISES 1(b)

In Nos. 1–8 prove the given statement by induction:

1. $1^3 + 2^3 + 3^3 + \cdots + n^3 = \frac{1}{4}n^2(n+1)^2$
2. $1^2 + 2^2 + 3^2 + \cdots + (2n-1)^2 = \frac{1}{3}n(4n^2 - 1)$
3. $1\cdot2 + 2\cdot3 + 3\cdot4 + \cdots + n(n+1) = \frac{1}{3}n(n+1)(n+2)$
4. $1 + 2x + 3x^2 + \cdots + nx^{n-1} = \dfrac{1 - (n+1)x^n + nx^{n+1}}{(1-x)^2} \quad (x \neq 1)$
5. $\sum_{r=1}^{n} (-1)^{r-1} r^2 = \frac{1}{2}(-1)^{n-1} n(n+1)$
6. $6^n - 5n + 4$ is divisible by 5
7. $9^{n+1} - 8n - 9$ is divisible by 64
8. $1 + 2\cos x + 2\cos 2x + \cdots + 2\cos nx = \sin(n + \frac{1}{2})x \operatorname{cosec} \frac{1}{2}x$ $(0 < x < 2\pi)$.
9. Given that $a_{r+1} = 2a_r + 1$ for $r = 1, 2, 3, \ldots$, and $a_1 = 1$, prove that $a_n = 2^n - 1$.
   Find the value of $\sum_{r=1}^{n} a_r$.

10. What is the fallacy in the following 'proof' that, given any set of n girls, if one is known to be blonde then all the girls are blonde?

Assume this is true for some particular n and consider any set of $n + 1$ girls of whom one (say B) is a blonde. Remove from the set one girl (say C) other than B. Then there remain n girls of whom at least one (B) is blonde, so by the inductive hypothesis all are blondes. Now put C back and remove another girl C'; now we have a set of n girls of whom $n - 1$ are known to be blondes, so the remaining girl is blonde. Hence all n are blondes.

The theorem is obviously true for $n = 1$ and so, by induction, is true for all positive integers n.

(I am indebted to Professor F.M. Arscott for this example.)

1.7 Manipulation of inequalities

Inequalities can be manipulated in almost the same way as equations, but there are important differences. If a and b are any real numbers, then $a \leqslant b$ means that either $a < b$ or $a = b$. Similarly $a \geqslant b$ means either $a > b$ or $a = b$. It follows that

$$\text{if } a \leqslant b \text{ and } a \geqslant b, \text{ then } a = b. \tag{1}$$

It is sometimes convenient to write $a \lessgtr b$ for $a < b$ or $a > b$. By definition $a > b$ means that $a - b > 0$ and $a < b$ means $b - a > 0$.

If a, b, c are any real numbers and $a < b$, then

$$a + c < b + c. \tag{2}$$

$$ac < bc \text{ if } c > 0$$

and

$$ac > bc \text{ if } c < 0. \tag{3}$$

Notice that Rule (2), which is equivalent to $(b + c) - (a + c) = b - a > 0$, allows us to transpose terms from one side of the inequality sign to the other. For example, $5 - 3x < x - 3$ is equivalent to $5 + 3 < x + 3x$, or $8 < 4x$; using Rule (3) this can be reduced to $2 < x$.

Rule (3) states $bc - ac = c(b - a)$ has the same sign as $b - a$ if c is positive and the opposite sign if c is negative.

Example 1 Find the values of x which satisfy $1/(x - 1) > -2$.

If $x - 1 > 0$, we have $1 > -2(x - 1)$ or $2x > 1$. That is $x > \frac{1}{2}$; since we assumed $x > 1$ $(x - 1 > 0)$, the original inequality holds for $x > 1$.

If $x - 1 < 0$, that is $x < 1$, we have $1 < -2(x - 1)$, by Rule (3), or

$x < \frac{1}{2}$. These two inequalities are simultaneously true only if $x < \frac{1}{2}$. Thus $1/(x-1) > -2$ if $x < \frac{1}{2}$ or if $x > 1$.

Example 2 Solve the inequality $x^2 - 5x + 6 < 0$.

Writing the quadratic in factor form we have

$$(x-2)(x-3) < 0,$$

and we see that the two factors must have opposite signs. Thus either

$$x-2 < 0 \text{ and } x-3 > 0 \quad \text{or} \quad x-2 > 0 \text{ and } x-3 < 0.$$

That is $x < 2$ and $x > 3$ or $x > 2$ and $x < 3$. The first of these pairs of inequalities is mutually inconsistent and so the required range of validity is $2 < x < 3$.

Successive applications of Rules (2) and (3) yield the following generalisations:

If $a_r < b_r$ for $r = 1, 2, 3, \ldots, n$, then

$$a_1 + a_2 + a_3 + \cdots + a_n < b_1 + b_2 + b_3 + \cdots + b_n \tag{4}$$

or

$$\sum_{r=1}^{n} a_r < \sum_{r=1}^{n} b_r .$$

If $0 < a_r < b_r$ for $r = 1, 2, 3, \ldots, n$, then

$$\prod_{r=1}^{n} a_r < \prod_{r=1}^{n} b_r . \tag{5}$$

Here $\prod_{r=1}^{n} a_r$ stands for the product $a_1 a_2 a_3 \cdots a_n$. From (5) we deduce that if $0 < a < b$, then

$$a^n < b^n \text{ for } n = 1, 2, 3, \ldots .$$

Example 3 For what values of x is

$$\frac{x-5}{5x-4} > \frac{x}{5x-3}?$$

This inequality is equivalent to

$$\frac{x-5}{5x-4} - \frac{x}{5x-3} > 0;$$

that is

$$\frac{5-8x}{(5x-4)(5x-3)} > 0.$$

Thus either
$$5-8x>0 \quad \text{and} \quad (5x-4)(5x-3)>0,$$
or
$$5-8x<0 \quad \text{and} \quad (5x-4)(5x-3)<0.$$

In the former case $x<\frac{5}{8}$ and $x<\frac{3}{5}$ or $x>\frac{4}{5}$; hence $x<\frac{3}{5}$.

The latter case yields $x>\frac{5}{8}$ and $\frac{3}{5}<x<\frac{4}{5}$; hence $\frac{5}{8}<x<\frac{4}{5}$. That is, the original inequality holds if
$$x<\tfrac{3}{5} \quad \text{or} \quad \tfrac{5}{8}<x<\tfrac{4}{5}.$$

Example 4 Prove that if a, b, c, d are real numbers such that $a^2+b^2=1$ and $c^2+d^2=1$, then $ac+bd \leqslant 1$.

We have $(a-c)^2 \geqslant 0$, and so $a^2+c^2 \geqslant 2ac$. Similarly $b^2+d^2 \geqslant 2bd$ and the result follows.

The *modulus* or *absolute value* of x is denoted by $|x|$, and is defined by
$$|x| = \begin{cases} x & \text{if } x \geqslant 0, \\ -x & \text{if } x<0. \end{cases}$$

Thus $|2|=2=|-2|$, $|0|=0$ and $|x| \leqslant 2$ means $-2 \leqslant x \leqslant 2$. Hence $|x-3|<4$ is equivalent to $-4<x-3<4$, or adding 3 to each term $-1<x<7$. Note that $-|x| \leqslant x \leqslant |x|$.

The *triangular inequality* is an important and frequently used result: for all values of a and b,
$$||a|-|b|| \leqslant |a \pm b| \leqslant |a|+|b|$$
Since
$$-|a| \leqslant a \leqslant |a| \text{ and } -|b| \leqslant b \leqslant |b|,$$
we have, by addition
$$-(|a|+|b|) \leqslant a+b \leqslant |a|+|b|.$$
That is
$$|a+b| \leqslant |a|+|b|. \tag{6}$$
Since
$$|-b| = |b|, |a-b| \leqslant |a|+|b|.$$

We can write $a=(a-b)+b$; then using the result just proved we have $|a| \leqslant |a-b|+|b|$, or
$$|a|-|b| \leqslant |a-b|.$$

Similarly, from $b=a+(b-a)$, we have
$$-|a-b| \leqslant |a|-|b|.$$
Hence
$$||a|-|b|| \leqslant |a-b|.$$

Clearly

$$||a| - |-b|| = ||a| - |b|| \leqslant |a+b|.$$

This result together with inequality (6) establishes the triangular inequality.

Example 5 Solve the inequality

$$\left|\frac{3x+4}{x-7}\right| < 1.$$

By squaring we can write the inequality

$$\left(\frac{3x+4}{x-7}\right)^2 < 1, \quad \text{or} \quad (3x+4)^2 < (x-7)^2.$$

That is $(4x-3)(2x+11) < 0$; hence $-\frac{11}{2} < x < \frac{3}{4}$.

Example 6 If $P(x) = x^m + a_1x^{m-1} + a_2x^{m-2} + \cdots + a_m$, where m is a positive integer and the a's are real numbers, then there is a real number X such that

$$\tfrac{1}{2}x^m \leqslant P(x) \leqslant \tfrac{3}{2}x^m \qquad \text{for } x > X.$$

Now

$$\begin{aligned} |a_1x^{m-1} + a_2x^{m-2} + \cdots + a_m| &\leqslant |a_1x^{m-1}| + |a_2x^{m-2}| + \cdots + |a_m| \\ &\leqslant Ax^{m-1} \qquad \text{if } x > 1, \end{aligned}$$

where $A = |a_1| + |a_2| + \cdots + |a_m|$.

Thus

$$|a_1x^{m-1} + a_2x^{m-2} + \cdots + a_m| \leqslant \tfrac{1}{2}x^m \qquad \text{if } x > X = \max(2A, 1).$$

That is

$$-\tfrac{1}{2}x^m \leqslant a_1x^{m-1} + a_2x^{m-2} + \cdots + a_m \leqslant \tfrac{1}{2}x^m,$$

and the result follows on the addition of x^m.

Another useful result is known as

Cauchy's inequality If $a_r, b_r, r = 1, 2, 3, \ldots, n$, are two sets of non-zero real numbers, then

$$\left(\sum_{r=1}^{n} a_rb_r\right)^2 \leqslant \left(\sum_{r=1}^{n} a_r^2\right)\left(\sum_{r=1}^{n} b_r^2\right),$$

and equality holds only when $a_r/b_r = c$ (constant $\neq 0$) for $r = 1, 2, 3, \ldots, n$.

Consider the quadratic in x:

$$P(x) = \sum_{r=1}^{n} (a_r x + b_r)^2 = x^2 \sum_{r=1}^{n} a_r^2 + 2x \sum_{r=1}^{n} a_r b_r + \sum_{r=1}^{n} b_r^2.$$

Since $P(x) \geqslant 0$ for all x and $\Sigma a_r^2 > 0$, the discriminant of $P(x)$ must be non-positive. That is

$$\left(\sum_{r=1}^{n} a_r b_r\right)^2 - \left(\sum_{r=1}^{n} a_r^2\right)\left(\sum_{r=1}^{n} b_r^2\right) \leqslant 0,$$

which is the required inequality.

Also $P(x) = 0$ only when a value of x exists such that

$$a_r x + b_r = 0, \qquad r = 1, 2, \ldots, n.$$

We also have

Minkowski's inequality If $a_r, b_r, r = 1, 2, \ldots, n$ are two sets of non-negative real numbers, then

$$\left\{\sum_{r=1}^{n} (a_r + b_r)^2\right\}^{1/2} \leqslant \left(\sum_{r=1}^{n} a_r^2\right)^{1/2} + \left(\sum_{r=1}^{n} b_r^2\right)^{1/2},$$

and equality holds only when $a_r/b_r = c$ ($\neq 0$) for all r.

$$\begin{aligned}\Sigma(a_r + b_r)^2 &= \Sigma a_r^2 + 2\Sigma a_r b_r + \Sigma b_r^2 \\ &\leqslant \Sigma a_r^2 + 2(\Sigma a_r^2)^{1/2} (\Sigma b_r^2)^{1/2} + \Sigma b_r^2,\end{aligned}$$

using Cauchy's inequality. That is

$$\Sigma(a_r + b_r)^2 \leqslant [(\Sigma a_r^2)^{1/2} + (\Sigma b_r^2)^{1/2}]^2,$$

and the result follows.

Example 7 If p is a real number, $p > -1$ ($\neq 0$) and n is a positive integer, then

$$(1 + p)^n > 1 + np, \qquad n = 2, 3, 4, \ldots .$$

(This result is known as *Bernoulli's inequality*.)

We use induction to prove this inequality. Suppose the inequality holds for some value of n, then

$$\begin{aligned}(1 + p)^{n+1} &> (1 + p)(1 + np) \\ &> 1 + (n + 1)p, \qquad \text{since } np^2 > 0.\end{aligned}$$

Also $(1+p)^2 = 1 + 2p + p^2 > 1 + 2p$; that is the inequality holds when $n = 2$, and the result follows by induction. Notice that Bernoulli's inequality can be written

$$a^n - 1 > n(a-1),$$

by writing $1 + p = a$.

EXERCISES 1(c)

In Nos. 1–14, find the values of x for which the given inequality is true:

1. $3x - 4 < 2 + x$
2. $x^2 - 1 \leqslant 3$
3. $x^2 - x - 6 < 0$
4. $x(1-x) < 1$
5. $x^2 + 3 \leqslant 2x$
6. $(x-1)^{-1} < (2-x)^{-1}$
7. $(x+5)/(2x-1) > 0$
8. $|x+2| < 3$
9. $0 < |x-4| < 3$
10. $|x+5| < 2|x|$
11. $\left|\dfrac{2-x}{x+3}\right| < 1$
12. $\dfrac{4x-3}{x-5} < 1$
13. $\dfrac{x+3}{x+1} < \dfrac{x-1}{x+3}$
14. $|x-7| < 1 < |x-1|$.

15. Show that if $2 \leqslant x \leqslant 4$ and $3 \leqslant y \leqslant 8$, then

    (a) $5 \leqslant x + y \leqslant 12$ (b) $-6 \leqslant x - y \leqslant 1$
    (c) $6 \leqslant xy \leqslant 32$ (d) $\frac{1}{4} \leqslant x/y \leqslant \frac{4}{3}$.

16. Show that if a, b, c, d are all positive and $a/b < c/d$, then

$$\frac{a}{b} < \frac{a+c}{b+d} < \frac{c}{d}.$$

17. Show that if $m/n > 0$ and $(m/n)^2 < 2$, then

$$\frac{3m+4n}{3n+2m} > \frac{m}{n} \text{ and } \left(\frac{3m+4n}{3n+2m}\right)^2 < 2.$$

18. Show that, if p/q is a positive rational approximation to $\sqrt{2}$, then $(p+2q)/(p+q)$ is a closer approximation and if the former is less than $\sqrt{2}$ then the latter is greater than $\sqrt{2}$.

19. Prove that if a, b, c, d are any real numbers, then

$$a^4 + b^4 \geqslant 2a^2b^2 \quad \text{and} \quad a^4 + b^4 + c^4 + d^4 \geqslant 4abcd.$$

    Prove also that

$$(a^2+b^2)^2 + (c^2+d^2)^2 \geqslant 2(ab+cd)^2.$$

    Show that if a, b, c, d are all positive and $a^4 + b^4 + c^4 + d^4 \leqslant 1$, then

$$a^{-4}+b^{-4}+c^{-4}+d^{-4}\geqslant 16.$$

20. Prove that if $\sum_{r=1}^{n} a_r = s$ and $a_r > 0, r = 1, 2, \ldots, n$, then

(a) $\prod_{r=1}^{n} (1+a_r) > 1+s$ (b) $\prod_{r=1}^{n} (1-a_r) > 1-s$ if $a_r < 1$.

21. Prove that if $a_r > 0, r = 1, 2, \ldots, n$ and $\sum_{r=1}^{n} a_r = s$, then

(a) $\prod_{r=1}^{n} (1+a_r) < \dfrac{1}{1-s}$, if $s < 1$

(b) $\prod_{r=1}^{n} (1-a_r) < \dfrac{1}{1+s}$.

CHAPTER 2

Functions and Limits

2.1 Functions

In more elementary texts a function is said to be the relation which enables one to determine the real number y when the real number x is given. Often the relation between the two variables may be expressed by means of formulae such as

$$y = x^2, \quad y = \sqrt{x}, \quad xy + \log y = \sin x.$$

Functions can also be represented by graphs; the temperature at a given place depends on time, that is it is a function of time, and can be indicated by the graph drawn by a self-recording thermometer.

Essentially a function is a set of ordered pairs. Let X and Y be two sets; if, to each x of X there is associated *one* element y of Y, then the relation is said to define a *function* f. We write $y = f(x)$ and call y the value of the function at x. Then f is the set of ordered pairs (x, y) or $(x, f(x))$ such that if (x, y_1) and (x, y_2) both belong to f, then $y_1 = y_2$.

The set X is called the *domain* and the set of values y the *range* of the function. The variable x is called the *argument* of the function; x and y are also called the *independent* and *dependent variables*. Notice that, to a given x there corresponds *just one* y, but the same y may correspond to more than one x. In the case of a constant the same y corresponds to every x in the domain of the function.

We also say that f is a *mapping of* X *into* Y, and that f *sends* x *to* y or y is the *image* of x under f. If the range of f concides with Y then the mapping of X is *onto* Y.

Example 1 $y = x^2$ is defined for all real numbers. Thus we may take for the domain the set of all real numbers. The range of this function is then the set of all real non-negative numbers. Note that to each value of x there corresponds one value of y, but that different values of x may lead to the same value of y; for example both $x = 2$ and $x = -2$ lead to $y = 4$.

Example 2 $y^2 = x$ does *not* determine a function since two distinct values $\pm y$ are assigned to each x. However $y = \sqrt{x}$ and $y = -\sqrt{x}$ each defines a function for $x \geqslant 0$.

Example 3 $y = \sqrt{(1 - x^2)}$ is defined only for $-1 \leqslant x \leqslant 1$; the range is $0 \leqslant y \leqslant 1$.

Example 4 y is the denominator of x when x is expressed in its lowest terms. This defines a function when x is a rational number. When $x = p/q$, where p and q are integers with no common divisor ($q > 0$), then $y = q$. The range of the function is the set of positive integers.

Example 5
$$y = \begin{cases} 0 & \text{when } x \text{ is rational,} \\ 1 & \text{when } x \text{ is irrational.} \end{cases}$$

Here the domain is the set of all real numbers and the range the numbers 0 and 1. Neither this nor the previous example can be represented graphically.

Example 6 $y = [x]$, that is y is the largest integer less than or equal to x. The domain is the set of real numbers; the range is the set of all integers. The graph is shown in Fig. 1; the left-hand end point (but not the right-hand one) of each step belongs to the graph. This function is called the *square-bracket function*.

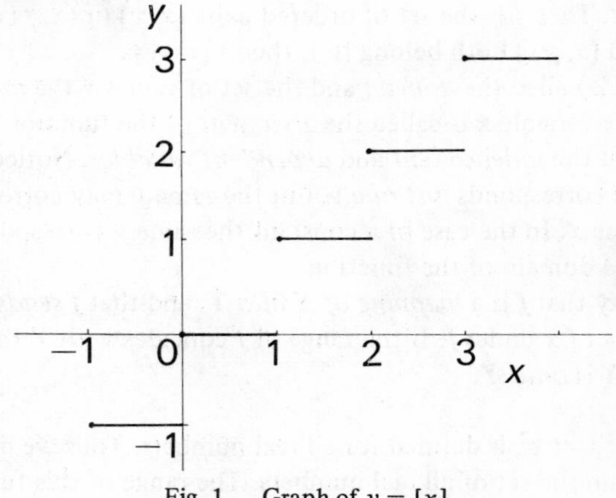

Fig. 1 Graph of $y = [x]$

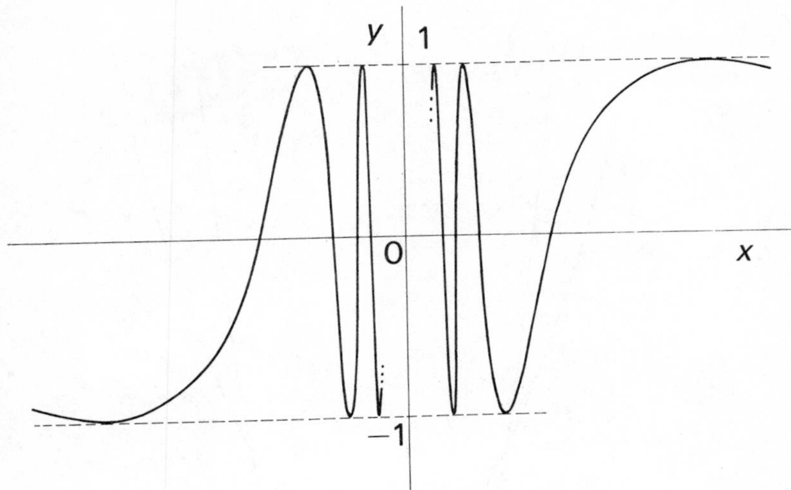

Fig. 2 Graph of $y = \sin(1/x)$

Example 7 *Heaviside's unit step function* $H(x - a)$ is defined to be zero when its argument is negative and unity when its argument is positive or zero. Thus

$$H(x-a) = \begin{cases} 0 \text{ when } x < a, \\ 1 \text{ when } x \geqslant a. \end{cases}$$

Notice that when $x > 0$ we can write

$$[x] = H(x-1) + H(x-2) + H(x-3) + \cdots .$$

Example 8 $y = \sin(1/x)$ is defined for all real x except $x = 0$. The graph of this function lies between the lines $y = \pm 1$.We have

$$y = 0 \qquad \text{when } x = 1/n\pi, \quad n = \pm 1, \pm 2, \ldots,$$

$$y = (-1)^{n-1} \qquad \text{when } x = \frac{2}{(2n-1)\pi}, \quad n = 0, \pm 1, \pm 2, \ldots .$$

The graph is shown in Fig. 2.

2.2 Types of function

A *polynomial* of degree n in x is a function of the form $a_0x^n + a_1x^{n-1} + a_2x^{n-2} + \cdots + a_n$, where $a_0, a_1, a_2, \ldots, a_n$ are all constants. The simplest polynomials are the powers $x, x^2, \ldots, x^n$. If graphs of these

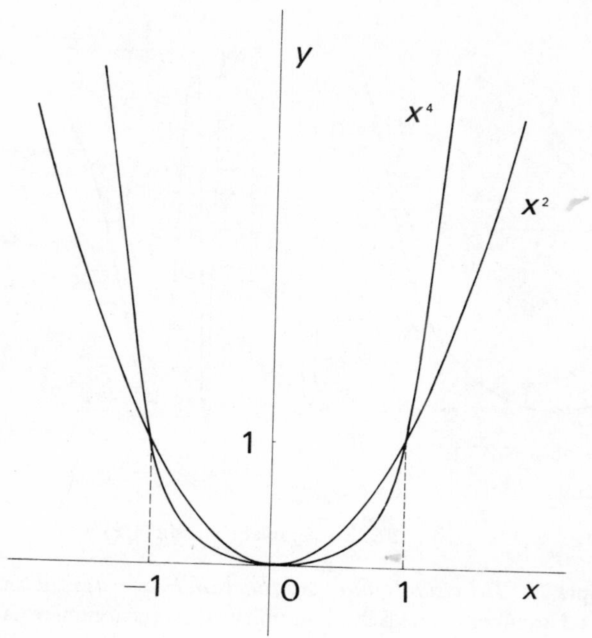

Fig. 3 Graph of $y = x^{2k}$, $k = 1, 2, \ldots$

powers are drawn it will be seen that the curves fall into two categories (Figs. 3, 4).

A function is said to be *even* if $f(-x) = f(x)$ and *odd* if $f(-x) = -f(x)$ for all x in the domain of the function. Thus a polynomial with only even powers of x is an even function and one containing only odd powers is odd. Other examples of even functions are $\cos x$, $x \sin x$; $\sin x$ and $x \cos x$ are odd functions.

A *rational function* is the quotient of two polynomials. In the particular case when the denominator is a constant the rational function reduces to a polynomial; hence the class of all polynomials is a sub-class of the class of all rational functions.

An *algebraic function* of x, of degree n, is the root of an equation of degree n in y whose coefficients are rational functions of x. For example $y = \sqrt{(x + \sqrt{x})}$ becomes $y^4 - 2xy^2 + x^2 - x = 0$ when the roots are removed by squaring. Sometimes it is not possible to solve for y explicitly, for example $y^5 + xy - x^3 = 0$. In this case the equation is said to define y as an *implicit function* of x.

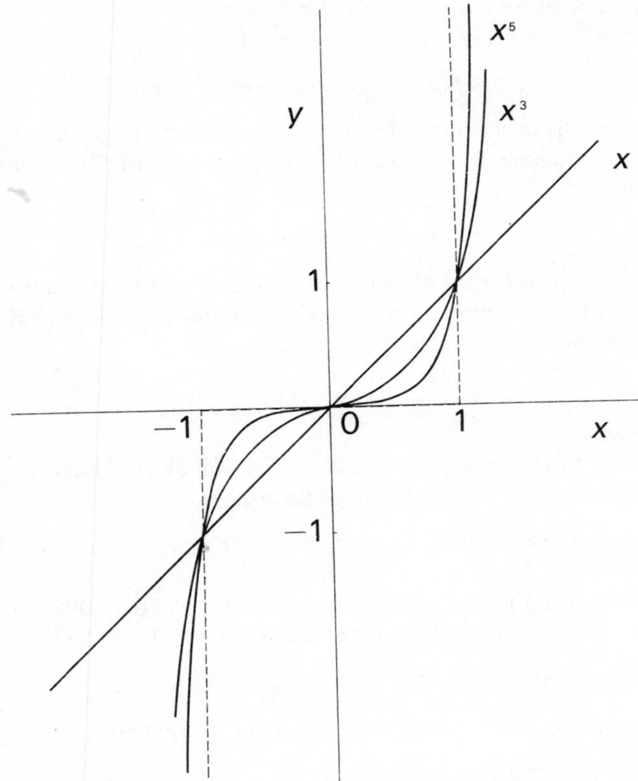

Fig. 4 Graph of $y = x^{2k-1}, k = 1, 2, 3, \ldots$

All functions which are not algebraic are called *transcendental*. This class of functions includes the trigonometric functions and their inverses, the logarithmic and exponential functions and the *hyperbolic functions*. The *hyperbolic cosine* and the *hyperbolic sine* are defined by the relations

$$\operatorname{ch} x = \tfrac{1}{2}(e^x + e^{-x}) \quad \text{and} \quad \operatorname{sh} x = \tfrac{1}{2}(e^x - e^{-x});$$

ch x is read 'cosh x' and sh x as 'shine x'. The ratio of these two functions is called the *hyperbolic tangent*,

$$\operatorname{th} x = \frac{\operatorname{sh} x}{\operatorname{ch} x},$$

and is read 'than x' as in 'thank'.*

We shall assume the transcendental functions for illustrative examples, assuming without proof that they possess the desired properties; we do not, however, use them to establish the properties we are illustrating.

* These functions are also written cosh x, sinh x and tanh x.

Some of the properties will be obtained in Chapter 8.

A function f is said to be *increasing* in an interval I if

$$f(x_1) \leqslant f(x_2) \qquad \text{whenever } x_1 < x_2,$$

and both points lie in the interval. An increasing function may also be described as *non-decreasing*. If equality is not allowed then f is *strictly increasing* in I. If

$$f(x_1) \geqslant f(x_2) \qquad \text{whenever } x_1 < x_2$$

and both points lie in I, then f is said to be *decreasing* in I. If equality is not allowed then f is called *strictly decreasing*. A function is *monotonic* if it is either increasing or decreasing throughout I.

EXERCISES 2(a)

These exercises introduce some functions which will be used later to illustrate the extent or limitations of the theory.

1. Compare the sizes of x^6, $1000x^3$, $1\,000\,000x$ when $x = 1, 10, 100, 1000$.

   Hence in a polynomial such as $x^6 + 1000x^3 + 1000000x$ the highest power of x overwhelms the other terms as x gets larger and larger.

2. Sketch the graphs of

   $$x \sin x, \quad x^{-1} \sin x, \quad x \sin(1/x), \quad x^2 \sin(1/x).$$

3. Sketch the graphs of

   $$x - [x], \quad (x - [x])^2, \quad \sqrt{(x - [x])}, \quad [x] + \sqrt{(x - [x])}.$$

   where $[x]$ is defined in Example 6, Section 2.1.

4. If there is a constant λ, such that $f(x + \lambda) = f(x)$ for all x in the domain of the function, then f is said to be *periodic* with period λ. For example, $\sin(x + 2\pi) = \sin x$ for all x; thus $\sin x$ has period 2π. Show that $\tan x$ has period π and $x - [x]$ has period 1.

5. Show that the periodic function f with period $2a$ and

   $$f(x) = \begin{cases} 1, & 0 < x < a, \\ 0, & a < x < 2a, \end{cases}$$

   can also be written in terms of the Heaviside unit step function

   $$f(x) = H(x) - H(x - a) + H(x - 2a) - H(x - 3a) + \cdots.$$

6. Express $x - [x]$ as (a) a periodic function, and (b) in terms of the Heaviside unit step function for $x > 0$.

7. Prove that a polynomial cannot be a periodic function.

8. From the definitions of the hyperbolic functions given in Section 2.2 prove the following identities:

   (a) $\text{ch}^2\, x - \text{sh}^2\, x = 1$ (b) $\text{ch}\,(x+y) = \text{ch}\, x\, \text{ch}\, y + \text{sh}\, x\, \text{sh}\, y$.

   (c) $\text{sh}\,(x+y) = \text{sh}\, x\, \text{ch}\, y + \text{ch}\, x\, \text{sh}\, y$

   (d) $\text{ch}\, 2x = \text{ch}^2 x + \text{sh}^2 x = 2\, \text{ch}^2\, x - 1 = 1 + 2\, \text{sh}^2\, x$

   (e) $\text{sh}\, 2\, x = 2\, \text{sh}\, x\, \text{ch}\, x$ (f) $\text{th}\, 2\, x = 2\, \text{th}\, x/(1 + \text{th}^2\, x)$.

9. Show that ch x is an even function of x and sh x is an odd function.

10. Show that the following functions are monotonic, that is, either increasing or decreasing functions:

    (a) $mx + c$ (b) $\sqrt{x}, \quad x \geqslant 0$

    (c) $x^{-1}, \quad x > 0$ (d) $\dfrac{x+1}{x-2}, \quad x > 2$

    (e) $\dfrac{x}{1+x^2}, \quad |x| < 1$ (f) $\dfrac{x}{1+x^2}, \quad x > 1$.

2.3 Bounds of a function

If there is a constant K such that $f(x) \leqslant K$ for all x in the domain of the function, then the function is said to be *bounded above*, and K is *an upper bound* of f. Graphically this means that no part of the curve $y = f(x)$ is above the line $y = x$. Similarly, if for a constant J, $f(x) \geqslant J$ for all x in the domain, then f is *bounded below* and J is *a lower bound*. A function which is bounded both above and below is said to be *bounded*. This implies that there is a constant M such that

$$|f(x)| \leqslant M \qquad \text{for all } x \text{ in the domain.}$$

A function which is not bounded is called *unbounded*.

Notice that to show a function f is bounded above it is necessary to find some value of K for which $f(x) \leqslant K$, but not necessarily the smallest such value. Thus $\sin x < 8$ for all x; indeed $\sin x \leqslant K$ for any $K \geqslant 1$.

Example 1 $f(x) = \sqrt{(1-x^2)}$, $|x| \leqslant 1$ is bounded since $|f(x)| \leqslant 1$ for $|x| \leqslant 1$.

Example 2 $f(x) = 1/x$, $0 < \delta \leqslant x \leqslant 1$ is bounded above as $f(x) \leqslant 1/\delta$ in this interval. However, if the domain is $0 < x < 1$, then the function is unbounded. In both cases the function is bounded below by unity.

2.4 Supremum and infimum

We have seen that if K is an upper bound of a given function f, then any number larger than K is also an upper bound. We now show that there must be a *least* value for the set of all possible K.

Theorem Let f be defined and bounded above in an interval I; then there exists a number M such that, for any $\epsilon > 0$,

$$\begin{aligned} f(x) &\leqslant M \quad \text{for every } x \text{ in } I, \text{ and} \\ &> M - \epsilon \quad \text{for at least one } x \text{ in } I. \end{aligned}$$

The number M is called the *supremum* or *the least upper bound* of f in I. We write the supremum as $\sup f(x)$ or $\sup f$, $\sup_{x \in I} f(x)$ or $\sup_I f$ if it is necessary to indicate the interval being considered.* Notice that the supremum need not be a value taken by the function (see Example 1 below).

This is one of the fundamental theorems in analysis and we prove it by means of a Dedekind section of the real numbers (see Section 1.4).

Proof Let Y, the range of f, be bounded above. Divide the real numbers r, into two classes L, R as follows:

$$r \in L \text{ if there is a } y \in Y \text{ such that } y > r,$$

$$r \in R \text{ if for all } y \in Y, y \leqslant r.$$

Neither class is empty; for R contains K where K is any upper bound of f, and, if y is any value taken by f, then $y-1$ or any smaller number must be in L. Also every real number must be in one or other class, and any member of L is less than all members of R.

Therefore, this section defines a real number M such that for all $\epsilon > 0$; $M - \epsilon \in L$ and $M + \epsilon \in R$. Suppose $M \in L$, then there exists $y \in Y$ such that $y > M$. Now the number $\eta = \frac{1}{2}(y + M)$ satisfies $y > \eta > M$ and $\eta \in R$ since $\eta > M$. But if $\eta \in R$, then $y \leqslant \eta$ by the second defining property, and this contradicts $y > \eta$. Therefore $M \in R$.

* The least upper bound is written lub f or $\operatorname{lub}_{x \in I} f(x)$.

That is, the set of all upper bounds has a *least* member.

Similarly, by reversing the inequalities, we obtain the *infimum* or *greatest lower bound* of a function which is bounded below.

The infimum is a number m such that, for any $\epsilon > 0$,

$$f(x) \begin{cases} \geqslant m & \text{for all } x \text{ in } I, \text{ and} \\ < m + \epsilon & \text{for at least one } x \text{ in } I. \end{cases}$$

We denote m by any of the following:

$$\inf f, \quad \inf_{x \in I} f(x), \quad \text{glb}\, f, \quad \text{glb}_{x \in I} f(x).$$

Clearly, for a bounded function

$$\inf f \leqslant \sup f.$$

Example 1 $f(x) = x^2, -1 < x < 2$ is bounded; $\inf f = 0$ and $\sup f = 4$.

Here f attains its infimum but not its supremum.

Example 2 $f(x) = -x^2$ is bounded above and $\sup f = 0$.

In this case f attains its supremum.

Example 3 $f(x) = \sin(1/x), 0 < x < \infty$ is bounded; $\inf f = -1$ and $\sup f = 1$ (See Example 8, Section 2.1).

Example 4 $f(x) = x^{-2}, 0 < x < 1$ is unbounded in any neighbourhood of the origin, so $\sup f$ does not exist; $\inf f = 1$.

EXERCISES 2(b)

1. Find $\sup f$ and $\inf f$, where they exist, for each of the following functions:

   (a) $f(x) = 1/x, -2 < x < -1$.

   (b) $f(x) = x^{-2}, 1 \leqslant x \leqslant 2$.

   (c) $f(x) = 1/x, 0 < |x| \leqslant 1; f(0) = 0$.

   (d) $f(x) = x \sin(1/x), x \neq 0; f(0) = 0, 0 < x < 2/\pi$.

   (e) $f(x) = x - [x], 0 < x < 10$.

2. Prove that if f is bounded on an interval $a \leqslant x \leqslant b$, then

$$\sup |f| - \inf |f| \leqslant \sup f - \inf f.$$

3. Prove that if f and g are both bounded on an interval $a \leqslant x \leqslant b$, then

   (a) $\inf f + \inf g \leqslant \inf (f+g) \leqslant \sup (f+g) \leqslant \sup f + \sup g$,

   (b) $\inf (f+g) \leqslant \inf f + \sup g \leqslant \sup (f+g)$.

4. The function f is defined for all real x and y, and satisfies the relation

$$f(x+y) = f(x)f(y).$$

   Show that if f is not identically zero, then f is positive for all x and $f(0) = 1$. Show also that f cannot be bounded.

2.5 Behaviour as $x \to \infty$

Let f be defined for all values of x greater than a fixed number c; that is f is defined for all sufficently large values of x. Suppose that as x increases indefinitely, $f(x)$ takes a succession of values which approach more and more closely the value l, and that the numerical difference between l and the values $f(x)$ taken by the function can be made as small as we please by taking values of x sufficiently large. Then we say that f *tends to the limit* l as x tends to infinity.

More precisely, $f(x)$ tends to l if, having chosen a positive number ϵ, we can then find a number X such that

$$|f(x) - l| < \epsilon \qquad \text{for all } x \geqslant X.$$

The number ϵ can be as small as we like; indeed the important criterion is that however small we take ϵ we can always find a value X for which the above inequality holds.

The definition can be expressed more succinctly if we introduce two symbols:

$\forall$ which means 'for all' or 'for every', and

$\exists$ which means 'there exists'.

We can then write

$$f(x) \to l \text{ as } x \to \infty \qquad \text{if } \forall \epsilon > 0; \exists X. \, |f(x) - l| < \epsilon \text{ for all } x \geqslant X.$$

In this form, what comes *before* the semicolon is chosen to begin with and what *follows* the semicolon depends on what precedes it. The full stop stands for 'such that'; some authors write this s.t. The number X depends on the choice of ϵ; when we wish to emphasise this we write $X(\epsilon)$.

Example 1 $f(x) = 1 - 1/x$ as $x \to \infty$

Here $|f(x) - 1| = 1/x$ (we have dropped the modulus sign since $x \to +\infty$ and so we can assume x to be positive), and $1/x < \epsilon$ if $x > 1/\epsilon$.
Thus we can take $X = \epsilon^{-1}$ for any given positive ϵ. To take a numerical illustration, if $\epsilon = 0{\cdot}001$, $1/x < 0{\cdot}001$ for $x > X = 10^3$. Similarly if $\epsilon = 10^{-6}$ then $X = 10^6$.

Example 2 $f(x) = (3x^2 + 1)/(x^2 - 3)$.

The function is defined for all real x except for $x = \pm\sqrt{3}$; so is certainly defined for sufficiently large x.

When x is large we can see that the function is roughly $3x^2/x^2 = 3$. We now *prove* that its limit is 3.

$$|f(x) - l| = \left|\frac{3x^2 + 1}{x^2 - 3} - 3\right| = \frac{10}{|x^2 - 3|} = \frac{10}{x^2 - 3} \qquad \text{if } x > \sqrt{3}.$$

Now $x^2 - 3 > x^2 - \frac{1}{2}x^2$ if $\frac{1}{2}x^2 > 3$.

Hence

$$|f(x) - l| < \frac{10}{\frac{1}{2}x^2} \quad \text{if } x > \sqrt{6}$$

$$< \epsilon \qquad \text{if } x^2 > 20\epsilon^{-1} \text{ and } x > \sqrt{6}.$$

Thus, if $x > X = \max(\sqrt{(20/\epsilon)}, \sqrt{6})$*, then $|f(x) - l| < \epsilon$, and so we have found an X fulfilling the conditions in the definition of a limit. Consequently

$$\lim_{x \to \infty} \frac{3x^2 + 1}{x^2 - 3} = 3.$$

Example 3 $f(x) = x^{-m}$, $m > 0$.

Since we are interested in the behaviour of this function for large values of x, we can assume $x > 1$. Let $x = 1 + p$, where $p > 0$. Then, using Bernoulli's inequality (Example 7, Section 1.7) we have

$$(1 + p)^m > 1 + pm > pm.$$

Hence

$$x^{-m} = (1 + p)^{-m} < (pm)^{-1} < \epsilon \qquad \text{if } p > 1/(m\epsilon),$$

that is if $x > X = 1 + 1/(m\epsilon)$. Therefore $\lim_{x \to \infty} x^{-m} = 0$ when $m > 0$.

* By max (a, b) we mean the larger of a and b; max (a, b, c, . . . , k) means the largest of a, b, c, . . . , k. Similarly, min (a, b) means the smaller of a, b; min (a, b, c, . . . , k) means the smallest of a, b, c, . . . , k.

Example 4 $$f(x) = \frac{2x^3 - 3x + 6}{x^3 - 5x + 1} \to 2 \text{ as } x \to \infty.$$

In this case

$$|f(x) - l| = \left|\frac{7x + 4}{x^3 - 5x + 1}\right| < \frac{7x + x}{|x^3 - 5x + 1|} \text{ if } x > 4,$$

$$< \frac{8x}{x^3 - 5x} = \frac{8}{x^2 - 5}$$

$$< \frac{8}{x^2 - \frac{1}{2}x^2} \quad \text{if } \tfrac{1}{2}x^2 > 5,$$

which is certainly true when $x > 4$. That is,

$$|f(x) - l| < 16/x^2 \quad \text{if } x > 4,$$

$$< \epsilon \qquad \text{if } x > X = \max(4\epsilon^{-1/2}, 4).$$

Therefore

$$\lim_{x \to \infty} \frac{2x^3 - 3x + 6}{x^3 - 5x + 1} = 2.$$

Note. We are not seeking the *least* value X for which $|f(x) - l| < \epsilon$, but only establishing the *existence* of *an* X. Thus in the above example we could write $x^2 - 5 > x^2 - \frac{1}{4}x^2$ if $\frac{1}{4}x^2 > 5$, which leads to $X' = \max(\sqrt{(32/3\epsilon)}, 2\sqrt{5})$.

We can write $|f(x) - l| < \epsilon$ in the definition of a limit without modulus signs:

$$-\epsilon < f(x) - l < \epsilon \quad \text{or} \quad l - \epsilon < f(x) < l + \epsilon.$$

From this we can deduce that *if f has a limit as $x \to \infty$ then f is bounded for all sufficiently large x.*

We do this by giving ϵ a definite positive value such as 1 or $\frac{1}{2}|l|$. Then since the existence of the limit implies that l is finite (and unique) we have, taking $\epsilon = 1$,

$$l - 1 < f(x) < l + 1 \qquad \text{for } x > X_1.$$

If we take $\epsilon = \frac{1}{2}|l|$, then we have

$$\tfrac{1}{2}|l| < |f(x)| < \tfrac{3}{2}|l| \qquad \text{for } x > X_2.$$

This last result shows that if $l > 0$, then

$$|f(x)| > 0 \qquad \text{for } x > X_2.$$

When the values taken by the function increase indefinitely with x, we say that $f(x)$ *tends to infinity as* $x \to \infty$. More precisely, $f(x) \to \infty$ as $x \to \infty$ if

$$\forall K > 0; \exists X.\ f(x) > K \text{ for all } x > X = X(K).$$

Similarly $f(x) \to -\infty$ as $x \to \infty$ if

$$\forall K > 0; \exists X.\ f(x) < -K \text{ for all } x > X.$$

Example 5 $f(x) = (x^2 - 3)/(x + 1)$.

Now $$x^2 - 3 > x^2 - \tfrac{1}{2}x^2 \text{ if } \tfrac{1}{2}x^2 > 3,$$

and $$x + 1 < x + x \text{ if } x > 1.$$

Then $$\frac{x^2 - 3}{x + 1} > \frac{\frac{1}{2}x^2}{2x} = \frac{x}{4} \text{ if } x > \sqrt{6},$$

$$> K \text{ if } x > X = \max(4K, \sqrt{6}).$$

Therefore $$\frac{x^2 - 3}{x + 1} \to \infty \text{ as } x \to \infty.$$

Example 6 If $f(x) \to l$ as $x \to \infty$ and $f(x) \leqslant K$ for all x, then

$$l \leqslant K.$$

Since $f(x) \to l$ as $x \to \infty$,

$$\forall \epsilon > 0; \exists X.\ |f(x) - l| < \epsilon \qquad \text{when } x > X.$$

That is $$l - \epsilon < f(x) < l + \epsilon \qquad \text{when } x > X.$$

Now suppose $l > K$; choose $\epsilon = l - K$. Then using the left-hand inequality, $l - (l - K) < f(x)$ when $x > X_1$. That is, $f(x) > K$ when $x > X_1$, contradicting the hypothesis that $f(x) \leqslant K$ for all x. Therefore $l \not> K$, which means $l \leqslant K$.

A function such as $\cos x$, although bounded, does not possess a limit as $x \to \infty$. In such a case the function is said to *oscillate* as $x \to \infty$. If the function is bounded it oscillates finitely, otherwise it oscillates infinitely. For example $\cos x$ and $x^2 \sin x$ oscillate finitely and infinitely respectively as $x \to \infty$.

If f is defined for sufficiently large negative x, then the behaviour of $f(x)$ as $x \to -\infty$ may be examined. The definitions are very similar to those previously considered. Thus,

$$f(x) \to l \text{ as } x \to -\infty \text{ if}$$
$$\forall \epsilon > 0; \exists X > 0. \ |f(x) - l| < \epsilon \text{ for all } x < -X.$$
$$f(x) \to \infty \text{ as } x \to -\infty \text{ if}$$
$$\forall K > 0; \exists X > 0. \ f(x) > K \text{ for all } x < -X.$$
$$f(x) \to -\infty \text{ as } x \to -\infty \text{ if}$$
$$\forall K > 0; \exists X > 0. \ f(x) < -K \text{ for all } x < -X.$$

EXERCISES 2(c)

1. Using the definition of limit prove the following:

(a) $\lim_{x \to \infty} (x^3 + 7)^{-1} = 0$ (b) $\lim_{x \to \infty} (x^2 - 2x - 1)^{-1} = 0$

(c) $\lim_{x \to \infty} \dfrac{4x^3 - 1}{3x^5 + 5} = 0$ (d) $\lim_{x \to \infty} \dfrac{3x^3 - 2x + 5}{4x^4 - x + 1} = 0$

(e) $\lim_{x \to \infty} \dfrac{3x^3 - 3x + 4}{2x^3 + 1} = \dfrac{3}{2}$ (f) $\lim_{x \to \infty} \left(\dfrac{3x}{x - 1} - \dfrac{2x}{x + 1} \right) = 1$

(g) $\lim_{x \to \infty} \dfrac{x^2 + 3}{x^2 - 5x - 3} = 1$ (h) $\lim_{x \to \infty} \dfrac{2x^3 - 3x - 4}{3x^3 - x^2 - 1} = \dfrac{2}{3}$

(i) $\lim_{x \to \infty} \dfrac{[x]}{x} = 1$ (j) $\lim_{x \to \infty} \dfrac{\sin x}{x} = 0$

(k) $\lim_{x \to -\infty} \dfrac{x^2 + 3}{x^2 - 5x - 3} = 1$ (l) $\lim_{x \to -\infty} \dfrac{\sin x}{x} = 0$

(m) $\dfrac{3x^5 + 5}{4x^3 - 1} \to \infty$ as $x \to \infty$ (n) $\dfrac{x^4}{2x^3 + 1} \to -\infty$ as $x \to -\infty$.

2. Prove that if

$$P(x) = x^m + a_1 x^{m-1} + a_2 x^{m-2} + \cdots + a_m,$$

where m is a positive integer and the a's real numbers, then there exists a number X such that, when $x > X$,

$$\tfrac{1}{2} x^m < P(x) < \tfrac{3}{2} x^m.$$

3. Prove that

$$\lim_{x \to \infty} \frac{a_0 x^m + a_1 x^{m-1} + \cdots + a_{m-1} x + a_m}{b_0 x^n + b_1 x^{n-1} + \cdots + b_{n-1} x + b_n} = 0 \text{ if } m < n;$$

prove also that the limit is infinite if $m > n$. What is the value of the limit when $m = n$?

4. Show that the signum function,

$$\operatorname{sgn} x = \begin{cases} -1 & \text{if } x < 0 \\ 1 & \text{if } x > 0 \\ 0 & \text{if } x = 0 \end{cases}$$

is represented by each of the following limits:

(a) $\lim\limits_{n \to \infty} \dfrac{nx}{1 + n|x|}$ (b) $\lim\limits_{n \to \infty} \operatorname{th}^{-1}(nx)$ (c) $\lim\limits_{n \to \infty} \dfrac{2}{\pi} \tan^{-1}(nx)$.

5. Prove that if $f(x) \to \infty$ as $x \to \infty$, then $1/f(x) \to 0$ as $x \to \infty$. Give an example which shows the converse is false.

2.6 Behaviour as $x \to a$

Let f be a function defined in a neighbourhood of $x = a$; then we can consider the behaviour of f as x approaches a. Suppose that as x approaches a the values taken by f approach more and more closely a value l. That is, suppose that the numerical difference between l and the values taken by f can be made as small as we please by taking values of x sufficiently close to a; then we say that f tends to the limit l as x tends to a. We write

$$\lim_{x \to a} f(x) = l \quad \text{or} \quad f(x) \to l \text{ as } x \to a.$$

Using the symbols introduced in Section 2.5, we can write

$$f(x) \to l \text{ as } x \to a \text{ if}$$

$$\forall \epsilon > 0; \exists \delta > 0. \ |f(x) - l| < \epsilon \qquad \text{for } 0 < |x - a| < \delta.$$

Notice that the inequality $|f(x) - l| < \epsilon$ must hold for all x in a suitable neighbourhood of a, but not necessarily at a itself; for the function may not be defined for $x = a$ (see Example 3 below). A neighbourhood without its centre is known as a *deleted neighbourhood* of a.

An immediate deduction from the definition of a limit is that if $\lim\limits_{x \to a} f(x) = l$, *then f is bounded in a deleted neighbourhood of a.* For when the limit exists

$$|f(x) - l| < \epsilon \qquad \text{for } 0 < |x - a| < \delta = \delta(\epsilon).$$

We can write the first inequality

$$l - \epsilon < f(x) < l + \epsilon,$$

and take $\epsilon = 1$ then

$$l-1<f(x)<l+1 \qquad \text{for } 0<|x-a|<\delta_1;$$

Or we can take $\epsilon=\frac{1}{2}|l|$, then if $l\neq 0$

$$\tfrac{1}{2}|l|<|f(x)|<\tfrac{3}{2}|l| \text{ for } 0<|x-a|<\delta_2.$$

The reader should formulate definitions for the other possible behaviour of f as $x\to a$. Thus

$$f(x)\to\infty \text{ as } x\to a \text{ if}$$

$$\forall K>0;\exists\,\delta.\,f(x)>K \text{ for } 0<|x-a|<\delta.$$

Example 1 $\lim_{x\to 2}(x^2-6x+10)=2.$

Here
$$\begin{aligned}|(x^2-6x+10)-2| &= |(x-2)^2-2(x-2)|\\ &\leqslant |x-2|^2+2|x-2|, \quad \text{using the triangular inequality, Section 1.7}\\ &< 3\delta \text{ if } |x-2|<\delta \text{ and } \delta<1.\end{aligned}$$

Now choose $\epsilon>0$, and we have

$$|(x^2-6x+10)-2|<\epsilon \quad \text{for } |x-2|<\delta = \min(\tfrac{1}{3}\epsilon, 1).$$

Example 2 $f(x)=\dfrac{2x^3-3x+4}{x^3+5x-1}\to -4$ as $x\to 0$.

Here

$$|f(x)-(-4)| = \frac{|6x^2+17|.|x|}{|x^3+5x-1|}.$$

Now $6x^2+17<23$ when $|x|<1$; $x^3+5x-1=0$ for some value of x between 0 and $\frac{1}{4}$, so we must ensure that this zero does not fall within the neighbourhood being considered. Thus if $|x|<1/10$ (say), $|x^3+5x-1|>\frac{1}{2}$, and

$$\begin{aligned}|f(x)+4| &< (23/\tfrac{1}{2})\,|x| \qquad \text{for } |x|<\tfrac{1}{10},\\ &<\epsilon \qquad\qquad\qquad \text{for } |x|<\epsilon/46.\end{aligned}$$

Hence we can take $\delta=\min(\frac{1}{10}, \epsilon/46)$ and the limit is established.

Example 3 $f(x)=(x^2-4)/(x-2)$.

Note that this function is not defined at $x=2$, and so $f(x)=x+2$ for all x *except* $x=2$. Clearly $f(x)$ is near 4 when x is near 2; we show that $f(x)\to 4$ as $x\to 2$.

$$\begin{aligned} |f(x)-4| &= |x-2|, & x\neq 2,\\ &< \epsilon & \text{for } 0<|x-2|<\delta = \epsilon. \end{aligned}$$

Example 4
$$f(x)=\frac{2x^2-3x+4}{x^3+5x+1}\to\frac{3}{7} \text{ as } x\to 1.$$

$$\begin{aligned} |f(x)-\tfrac{3}{7}| &= \frac{|11x^3-36x+25|}{|x^3+5x+1|}\\ &= \frac{|11(x-1)^3+33(x-1)^2-3(x-1)|}{|x^3+5x+1|}\\ &\leqslant \frac{|x-1|(11|x-1|^2+33|x-1|+3)}{|x^3+5x+1|}\\ &< \frac{47|x-1|}{|x^3+5x+1|} \qquad \text{if } |x-1|<1. \end{aligned}$$

When $|x-1|<1$, that is $0<x<2$, the denominator $x^3+5x+1>1$. Thus

$$\begin{aligned} |f(x)-\tfrac{3}{7}| &< 47|x-1| & \text{for } |x-1|<1,\\ &< \epsilon & \text{for } |x-1|<\delta=\min(1, \epsilon/47). \end{aligned}$$

EXERCISES 2(d)

1. Using the definition of limit prove the following:

(a) $\lim\limits_{x\to 2}(x^2-2x-1)=-1$ (b) $\lim\limits_{x\to 1}(x^3-3x^2+5x+4)=7$

(c) $\lim\limits_{x\to -2}(x^2+3x+4)=2$ (d) $\lim\limits_{x\to 0} x\sin(1/x)=0$

(e) $\lim\limits_{x\to 0}\dfrac{4x^2-1}{3x^2+5}=-\dfrac{1}{5}$ (f) $\lim\limits_{x\to 1}\dfrac{3x^3-3x+4}{2x^3+1}=\dfrac{4}{3}$

(g) $\lim\limits_{x\to 1}\dfrac{x^2-3x+2}{x^3-3x^2+2}=\dfrac{1}{3}$ (h) $\lim\limits_{x\to -4}\dfrac{x^2+5x+4}{x^2+3x+2}=0.$

2. Prove that, if $\lim\limits_{x\to a} f(x)$ exists, then it is unique.

3. Prove that if $\lim\limits_{x\to a} f(x)=l$ and $l>0$, then $f(x)>0$ in a neighbourhood of $x=a$.

4. Prove that if $f(x)\to l$ as $x\to a$ and $f(x)>0$ in a neighbourhood of $x=a$, then $l\geqslant 0$.

5. Prove that if, as $x \to a, f(x) \to l, g(x) \to L$ and $f(x) < g(x)$ in a neighbourhood of $x = a$, then $l \leqslant L$.

6. Prove that if $f(x) \leqslant g(x) \leqslant h(x)$ in some neighbourhood of $x = a$, and as $x \to a$ both $f(x)$ and $h(x)$ have the limit l, then $g(x) \to l$ as $x \to a$.

7. Prove that, if k is a positive constant and $f(x) \to l$ as $x \to a$, then $kf(x) \to kl$ as $x \to a$.

8. Prove that, if $f(x) \to l$ as $x \to a$, then $|f(x)| \to |l|$ as $x \to a$. Is the converse true?

9. Show that if, for $x > a$, f is an increasing function and $\sup_{x > a} f(x) = M$ then $\lim_{x \to \infty} f(x) = M$.

2.7 One-sided limits

If, in the definition of a limit, x is restricted to values greater than a, we say that x approaches a *from the right* or *from above* and write $x \to a + 0$ or simply $x \to a+$. This leads to the *right-hand limit* or *limit from above*

$$\lim_{x \to a+} f(x) \quad \text{or} \quad f(a+0).$$

More precisely,

$$f(a+0) = l \text{ if}$$

$$\forall \epsilon > 0; \exists \delta > 0. \; |f(x) - l| < \epsilon \qquad \text{for } a < x < a + \delta.$$

Similarly, if the approach to a is through values less than a, we say that x approaches a *from the left* or *from below* and write

$$x \to a - 0 \quad \text{or} \quad x \to a-.$$

The *left-hand limit* or *limit from below* is defined to be l if

$$\forall \epsilon > 0; \exists \delta > 0. \; |f(x) - l| < \epsilon \qquad \text{for } a - \delta < x < a.$$

We write

$$\lim_{x \to a-} f(x) = l = f(a-0).$$

For example,

$$\lim_{x \to 0-} \frac{|x|}{x} = -1 \quad \text{and} \quad \lim_{x \to 0+} \frac{|x|}{x} = +1.$$

Note that $\lim_{x \to a} f(x) = l$ if and only if

$$f(a-0) = l = f(a+0).$$

Example If $f(x) = x^k$, where k is a fixed positive integer, then $f(1-0) = 1$.

Since $x < 1$, $1 - x^k > 0$ and

$$\begin{aligned} 1 - x^k &= (1-x)(1 + x + x^2 + \cdots + x^{k-1}) \\ &< k(1-x), \qquad \text{for } 0 < x < 1, \\ &< \epsilon \qquad \text{for } 0 < 1 - x < \delta = \epsilon/k. \end{aligned}$$

That is

$$|f(x) - 1| < \epsilon \qquad \text{for } 0 < 1 - \delta < x < 1.$$

2.8 Limit theorems

It would be tedious if we had to evaluate all limits from first principles as in Section 2.5 and 2.6. Indeed it seems 'natural' to say in Example 3, Section 2.5

$$\lim_{x \to \infty} \frac{2x^3 - 3x + 6}{x^3 - 5x + 1} = \lim_{x \to \infty} \frac{2 - \dfrac{3}{x^2} + \dfrac{6}{x^3}}{1 - \dfrac{5}{x^2} + \dfrac{1}{x^3}} = 2,$$

since $1/x^n \to 0$ as $x \to \infty$, when n is a positive integer. This, however, assumes the results of the theorem we are about to prove.

Theorem If $f(x) \to \alpha$ and $g(x) \to \beta$ as $x \to a$, then

(a) $f(x) \pm g(x) \to \alpha \pm \beta$, as $x \to a$;

(b) $f(x)\, g(x) \to \alpha\beta$ as $x \to a$;

(c) $f(x)/g(x) \to \alpha/\beta$ as $x \to a$, provided $\beta \neq 0$.

Note. The theorem remains true if a is replaced by $+\infty$ or $-\infty$.

Proof. Since f and g have limits α, β as $x \to a$, then $\forall \epsilon > 0; \exists\, \delta_1$ and δ_2 such that

$$|f(x) - \alpha| < \epsilon \text{ for } 0 < |x - a| < \delta_1,$$

and

$$|g(x) - \beta| < \epsilon \text{ for } 0 < |x - a| < \delta_2.$$

Hence

$$\begin{aligned} |\{f(x) \pm g(x)\} - \{\alpha \pm \beta\}| &< |f(x) - \alpha| + |g(x) - \beta| \\ &< 2\epsilon \qquad \text{for } 0 < |x - a| < \delta, \end{aligned}$$

where $\delta = \min(\delta_1, \delta_2)$.

Since ϵ is any positive number as small as we please, then 2ϵ serves just as well as ϵ; indeed we could take $k\epsilon$, where k is any fixed positive number. This proves (a).

Again, $\forall \epsilon > 0; \exists \delta_1$ and δ_2 such that

$$|f(x) - \alpha| < k\epsilon \quad \text{for } 0 < |x - a| < \delta_1,$$

and

$$|g(x) - \beta| < k\epsilon \quad \text{for } 0 < |x - a| < \delta_2,$$

where k is a definite positive number to be assigned later.

Also, choosing $k\epsilon = \frac{1}{2}|\beta| > 0$, we have

$$\tfrac{1}{2}|\beta| < |g(x)| < \tfrac{3}{2}|\beta| \qquad \text{for } 0 < |x - a| < \delta_3 \qquad \text{(see Section 2.6).}$$

Then

$$\begin{aligned} |f(x)\,g(x) - \alpha\beta| &\leqslant |g(x)|\,|f(x) - \alpha| + |\alpha|\,|g(x) - \beta| \\ &< (\tfrac{3}{2}|\beta| + |\alpha|)k\epsilon \text{ for } 0 < |x - a| < \delta, \end{aligned}$$

where $\delta = \min(\delta_1, \delta_2, \delta_3)$.

Now if we choose $k < (\frac{3}{2}|\beta| + |\alpha|)^{-1}$,

$$|f(x)\,g(x) - \alpha\beta| < \epsilon \qquad \text{for } 0 < |x - a| < \delta,$$

and we have proved (b).

Again,

$$\begin{aligned} \left|\frac{f(x)}{g(x)} - \frac{\alpha}{\beta}\right| &\leqslant \frac{|\beta|\,|f(x) - \alpha| + |\alpha|\,|g(x) - \beta|}{|\beta|\,|g(x)|} \\ &< \frac{(|\beta| + |\alpha|)k\epsilon}{\frac{1}{2}|\beta|^2} \qquad \text{for } 0 < |x - a| < \delta, \\ &< \epsilon, \qquad \text{choosing } k < 2|\beta|^2/(|\alpha| + |\beta|). \end{aligned}$$

This proves (c).

EXERCISES 2(e)

1. Find the value of each of the following one-sided limits:

(a) $\lim_{x \to 3-} (x - [x])$ (b) $\lim_{x \to 1+} ([x] + 1)^{-1}$

(c) $\lim_{x \to 1-} ([x] - 1)^{-1}$ (d) $\lim_{x \to \pi/2+} \tan x$

(e) $\lim_{x \to 0-} \cot x$ (f) $\lim_{x \to 4+} [\sqrt{x}]$

(g) $\lim_{x \to 4-} [\sqrt{x}]$ (h) $\lim_{x \to 2-} [x + (x - [x])^2]$

2. The function f is defined as follows:

$$f(x) = -x^2, \quad x \leqslant 0; \qquad f(x) = 5x - 4, \quad 0 < x \leqslant 1;$$
$$f(x) = [x], \quad 1 < x \leqslant 4; \qquad f(x) = x^2 - 8x + 19, \quad x > 4.$$

Find the left- and right-hand limits at $x = 0, 1, 2, 3, 4, 5$.

3. Prove that if, as $x \to \infty$, $f(x) \to \alpha$, $g(x) \to \beta$, then as $x \to \infty$,

(a) $f \pm g \to \alpha \pm \beta$,

(b) $fg \to \alpha\beta$,

(c) $f/g \to \alpha/\beta$, $\beta \neq 0$.

4. Using the limit theorems, find the values of the following limits:

(a) $\lim_{x \to 0} \dfrac{x^3 - 1}{x^2 + 1}$ (b) $\lim_{x \to \infty} \dfrac{x^2 - 1}{x^3 + 1}$

(c) $\lim_{x \to \infty} \dfrac{1 - x^2 + x^3}{1 + 2x^2 + 3x^3}$ (d) $\lim_{x \to 0} \dfrac{5x^2 + 2}{2x^2 - 3x + 1}$

(e) $\lim_{x \to \infty} \dfrac{5x^2 + 2}{2x^2 - 3x + 1}$ (f) $\lim_{x \to \infty} \dfrac{\operatorname{th} x}{x}$

5. Prove that, if n is a positive integer, then

$$\lim_{x \to a} \frac{x^n - a^n}{x - a} = na^{n-1}.$$

Using the limit theorems deduce that this result holds for any rational n.

6. Show that $\operatorname{th} x \to \pm 1$ as $x \to \pm\infty$.

CHAPTER 3

Continuity

3.1 Continuity

We think of a continuous curve as being one which we can drawn without taking our pen off the paper. The function of Example 8, Section 2.1 has a continuous graph except at $x = 0$, whereas the graph of Example 6, Section 2.1 has a break at every integer n. We are led to the conclusion that, for a function f to be continuous at $x = a$, it is necessary that $f(a)$ is defined and finite and also that the values taken by f must approach the value $f(a)$ as x approaches a from either side.

Formally, we have the definition: f is *continuous* at $x = a$ if

$$\forall \epsilon > 0; \exists \delta > 0. \ |f(x) - f(a)| < \epsilon \qquad \text{for } |x - a| < \delta.$$

A comparison with the definition in Section 2.6 shows that this is equivalent to saying

$$\lim_{x \to a} f(x) = f(a).$$

Also, from Section 2.7 we see that f is continuous at $x = a$ if $f(a - 0)$, $f(a)$, and $f(a + 0)$ all exist and are equal.

If $f(a)$ exists and the left-hand limit of $f(x), f(a - 0)$, is $f(a)$ then we say that f is *continuous on the left* at $x = a$. More precisely, if

$$\forall \epsilon > 0; \exists \delta > 0. \ |f(x) - f(a)| < \epsilon \qquad \text{for } a - \delta < x \leqslant a,$$

then f is continuous on the left at $x = a$.

A similar definition can be given for *continuity on the right* at $x = a$ when $f(a + 0) = f(a)$;

$$\forall \epsilon > 0; \exists \delta > 0. \ |f(x) - f(a)| < \epsilon \qquad \text{for } a \leqslant x < a + \delta.$$

Note. f is continuous at $x = a$ if and only if it is continuous on the left and on the right at $x = a$.

Example 1 $f(x) = x^3$ is continuous at $x = a$, for any finite a, since

$$|f(x) - f(a)| = |x^3 - a^3| = |x - a|\,|x^2 + ax + a^2|.$$

Now

$$|x^2 + ax + a^2| < |x|^2 + |a|\,|x| + |a|^2, \\ < 3(1 + |a| + |a|^2) \qquad \text{if } |x| < 1 + |a|.$$

Hence

$$|f(x) - f(a)| < \epsilon \qquad \text{when } |x - a| < \delta,$$

where

$$\delta = \min\{1, \tfrac{1}{3}\epsilon/(1 + |a| + |a|^2)\}.$$

Example 2 $f(x) = x^2, x \neq 1; f(1) = 0$ is not continuous at $x = 1$, since, as $x \to 1$, f approaches a value different from $f(1)$.

Example 3 $f(x) = x \sin(1/x), x \neq 0; f(0) = 0$ is continuous at $x = 0$, since

$$|f(x) - f(0)| = |x \sin(1/x)| \leqslant |x| \qquad \text{for all } x \neq 0, \\ < \epsilon \qquad \text{when } |x| < \delta = \epsilon.$$

3.2 Discontinuities

A point at which f is not continuous is said to be a *discontinuity* of f. There are four types:

(a) A *simple discontinuity* is one in which either $f(a)$ is not defined or, if it is defined, $f(a)$ has a value different from $\lim_{x \to a} f(x)$.

For example, $x \sin(1/x)$ is not defined at $x = 0$. Also $[1 - x^2] \to 0$ as $x \to 0$ from the left and from the right, but the function takes the value 1 at the origin. This type of discontinuity is *removable* by a suitable definition of $f(a)$.

(b) A *jump discontinuity* is one for which $f(a - 0) \neq f(a + 0)$. For example, $|x|/x$ at $x = 0$.

(c) If f is unbounded in any neighbourhood of $x = a$, then $x = a$ is said to be an *infinity* of f. For example, $(x - a)^{-2}$ at $x = a$.

(d) A discontinuity which is not one of the above types is called an *oscillatory* discontinuity. For example, $\sin(1/x)$ at $x = 0$.

3.3 Continuity in an interval

A function f is said to be continuous in the open interval (a, b) if it is continuous at each point of the interval. In the case of a closed interval $[a, b]$ values of x outside the interval must be avoided, and so a and b can be approached only through points within the interval. Consequently, we say f is continuous in the closed interval $[a, b]$ if f is continuous in the open interval (a, b), and

$$f(a+0) = f(a), \quad f(b-0) = f(b).$$

A function which possesses a finite number of removable or jump discontinuities in an interval is called *sectionally continuous*. For such a function $f(x-0)$ and $f(x+0)$ both exist for all x in the open interval, and are equal at all except a finite number of points. If the interval $[a, b]$ is closed, then both $f(a+0)$ and $f(b-0)$ must exist.

EXERCISES 3(a)

Use the ϵ–δ definition of continuity to prove that the functions in Nos. 1–6 are continuous at the points stated:

1. x^2 at $x = 2$
2. x^{-2} at $x = 1$
3. $|x|$ at $x = -1$
4. $\sqrt{x}$ at $x = 2$
5. x^4 at $x = a$
6. $(x^2 - 4)/(x - 1)$ at $x = 0$ and $x = 2$.

For what values of x in its domain of definition is each of the functions in Nos. 7–10 continuous?

7. $x^2/(x-2)$
8. $[x] - [-x]$
9. $\sqrt{[(x-a)(b-x)]}$, $b > a$
10. $\sqrt{[(x-a)/(b-x)]}$, $b > a$
11. Prove that, if f is continuous at $x = a$ and $f(a) > 0$, then there is a $\delta > 0$ such that $f(x) > 0$ for $a - \delta < x < a + \delta$.
12. Prove that, if f is continuous at $x = a$, then there is a $\delta > 0$ such that f is bounded in $a - \delta < x < a + \delta$.
13. Show that, in $[0, 1]$, the function

$$f(x) = \begin{cases} x, & \text{when } x \text{ is rational,} \\ 1 - x, & \text{when } x \text{ is irrational,} \end{cases}$$

is continuous only at $x = \frac{1}{2}$.

14. Given

$$f(x) = \begin{cases} 1/q, \text{ when } x = p/q, \\ 0 \text{ when } x \text{ is irrational,} \end{cases}$$

show that f is continuous at each irrational value of x in $(0, 1)$.

3.4 Continuity theorems

The limit theorems of Section 2.8 have counterparts in terms of continuity.

Theorem If f and g are both continuous at $x = a$, then so also are $f + g$, fg and f/g, provided that $g(a) \neq 0$.

The proofs of these properties follow closely the arguments used in Section 2.8, and are left as exercises.

Since $x^n = xx^{n-1}$, when n is a positive integer, we can use induction and the theorem on the product of continuous functions to prove x^n is continuous for all finite values of x. The quotient of continuous functions then establishes the continuity of x^{-n} for all non-zero values of x.

Further use of the above theorems shows that any *polynomial* is continuous for all finite values of x. Also a *rational function* is continuous for all finite values of x, except those making the denominator zero.

Theorem A continuous function of a continuous function is continuous. That is, if $t = g(x)$ is continuous at $x = a$, $f(t)$ continuous at $t = \alpha$ and $g(a) = \alpha$, then $f(g(x))$ is continuous at $x = a$.

Proof. Since f is continuous at $t = \alpha$, for any $\epsilon > 0$ there is an $\eta > 0$ such that

$$|f(t) - f(\alpha)| < \epsilon \qquad \text{for } |t - \alpha| < \eta.$$

Further, since g is continuous at $x = a$, there is a $\delta > 0$ such that

$$|g(x) - g(a)| < \eta \qquad \text{for } |x - a| < \delta.$$

Combining these inequalities yields

$$|f(g(x)) - f(g(a))| = |f(t) - f(\alpha)| < \epsilon \qquad \text{for } |x - a| < \delta,$$

which proves the theorem.

3.5 Functions continuous in an interval

Theorem If f is continuous in the closed interval $[a, b]$, then f is bounded in $[a, b]$.

Proof. Since f is continuous on the right at $x = a$ we know that f is bounded in $a \leqslant x < a + \delta$, for a suitable $\delta > 0$ (Exercises 3(a), No. 12). Let B be the set of points $x \in [a, b]$ for which f is bounded in $[a, x]$. Then B is not empty and all members of B are less than or equal to b. Consequently B possesses a supremum ξ, $a < \xi \leqslant b$.

Suppose $a < \xi < b$; then since f is continuous at ξ,

$$|f(x) - f(\xi)| < \tfrac{1}{2} \qquad \text{for } |x - \xi| < \delta_1,$$

(choosing $\epsilon = \frac{1}{2}$ in the definition of continuity). That is

$$f(x) < f(\xi) + \tfrac{1}{2} \qquad \text{for } \xi - \delta_1 < x < \xi + \delta_1;$$

but this implies $\xi + \delta_1 \in B$, which is impossible since ξ is the supremum of B. Therefore $\xi = b$ and the theorem is proved. (See also Section 5.9.)

Notice that it is essential for the interval of continuity of f to be closed. For example, $f(x) = 1/x$ is continuous in $0 < x < 1$, but is unbounded at the origin.

Theorem If f is continuous in the closed interval $[a, b]$, then it attains its bounds at points of the interval.

That is, if $m = \inf f$ and $M = \sup f$ for x in $[a, b]$, then there are points x_1, x_2 of $[a, b]$ at which

$$f(x_1) = m \quad \text{and} \quad f(x_2) = M.$$

Proof. As in the previous theorem it is essential that the interval should be closed. For example, $f(x) = x$ in the open interval $0 < x < 1$ is bounded ($m = 0, M = 1$) but does not take either of these values at points of the interval. Notice also that the points x_1, x_2 are not necessarily unique; thus the function x^2 attains its supremum at two points ± 1 in the interval $-1 \leqslant x \leqslant 1$.

We prove the theorem for upper bounds, leaving the case of lower bounds as an exercise. The existence of m, M is established by the previous theorem. Suppose there is no x in $[a, b]$ at which $f(x) = M$, then the function

$$\phi(x) = [M - f(x)]^{-1} \text{ is continuous in } [a, b]$$

and so is bounded.

By the properties of the supremum, there exists a point $x' \in [a, b]$ such that $f(x') > M - \epsilon$ for any chosen $\epsilon > 0$; but this implies that $\phi(x') > 1/\epsilon$ is unbounded. Therefore the denominator must vanish.

Theorem If f is continuous in a closed interval $[a, b]$ and $f(a) < 0$, $f(b) > 0$, then there is at least one point ξ, $a < \xi < b$, at which $f(\xi) = 0$.

Proof. This theorem asserts that if the graph of a continuous function starts below the x-axis and finishes above it, then the curve must cut the x-axis at least once. Although this is intuitively obvious, it is necessary to supply an analytical proof. As in the first theorem of the section we use the existence of the supremum of a bounded set.

Let A be the set of points of $[a, b]$ at which $f(x) < 0$. This set contains a and so is not empty. It possesses, therefore, a supremum ξ. Also, since $f(a) < 0$, there is a $\delta_1 > 0$ such that $f(x) < 0$ for $a \leqslant x < \delta_1$ (Exercise 3(a), No. 11), and so $\delta_1 \leqslant \xi$. Similarly $f(x) > 0$ for $\delta_2 < x \leqslant b$; these points are not in A, so $a < \delta_1 \leqslant \xi \leqslant \delta_2 < b$.

We now show that $f(\xi)$ is neither negative nor positive. For if $f(\xi) < 0$, then since f is continuous at ξ, $f(x) < 0$ in $\xi - \delta < x < \xi + \delta$ for some $\delta > 0$; in particular, $f(x) < 0$ for $\xi < x < \xi + \delta$. This contradicts the fact that ξ is the supremum of the set of x in $[a, b]$ for which f takes negative values. Therefore $f(\xi) \nless 0$.

Similarly $f(\xi) > 0$ leads to a contradiction and so $f(\xi) \ngtr 0$. Therefore $f(\xi) = 0$, and the theorem is proved.

Another way of expressing this theorem is known as the *intermediate value theorem*:

Theorem If f is continuous in a closed interval $[a, b]$ and $f(a) = \alpha$, $f(b) = \beta$ $(\alpha < \beta)$, then for any μ, $\alpha < \mu < \beta$, there exists a point ξ, $a < \xi < b$, such that $f(\xi) = \mu$.

Proof. This follows directly by applying the previous theorem to the function $g(x) = f(x) - \mu$.

In common with the previous theorems of this section it is essential for the interval of continuity of f to be closed. For example $f(x) = [x]$, $0 \leqslant x \leqslant 1$ does not take the value $\frac{1}{2}$ (say) in this interval.

EXERCISES 3(b)

1. Prove that if f and g are both continuous at $x = a$, then $f \pm g$ and fg are continuous at $x = a$.

2. Prove that if f and g are both continuous at $x = a$, and $g(a) \neq 0$, then f/g is continuous at $x = a$.

3. Show that if

$$f(x) = \sin x \quad \text{and} \quad g(x) = \begin{cases} x - \pi, & x \leqslant 0 \\ x + \pi, & x > 0, \end{cases}$$

then $f(g(x))$ is continuous at $x = 0$, but $g(x)$ is not continuous at $x = 0$.

4. Show that if

$$f(x) = \lim_{n \to \infty} \frac{\log(x+2) - x^{2n} \sin x}{1 + x^{2n}} \qquad \text{for } 0 \leqslant x \leqslant \tfrac{1}{2}\pi,$$

then f does not vanish anywhere in this interval, although $f(0)$ and $f(\frac{1}{2}\pi)$ are of opposite signs. Sketch the graph of f in the interval.

5. Show that, if f and g are continuous at $x = a$, so also are $\max(f, g)$ and $\min(f, g)$, where

$$\max(f, g) = \tfrac{1}{2}\{f + g + |f - g|\}, \quad \min(f, g) = \tfrac{1}{2}\{f + g - |f - g|\}.$$

6. Prove that if f is monotonic increasing in $[a, b]$ and takes all values between $f(a)$ and $f(b)$, then f is continuous in $[a, b]$.

7. Show that if f is continuous in $[a, b]$ and $K = f(c) + f(d)$, where $a < c < d < b$, then there is a point ξ, $a < \xi < b$, such that $f(\xi) = \frac{1}{2}K$.

8. Show that the function defined in Exercises 3(a), No. 13 takes all values between 0 and 1 in the interval $0 \leqslant x \leqslant 1$.

3.6 Inverse functions

A function f is said to be *one-to-one* if $f(x_1) = f(x_2)$ implies that $x_1 = x_2$. In this case there is a function g whose domain is the range of f and which is such that if f sends x to y, then g sends y back to x. Such a function is called the *inverse* of f and is usually written f^{-1}.

Thus, if $y = f(x)$ then $x = f^{-1}(y)$; Also

$$f(f^{-1}(y)) = y \text{ for all } y \text{ in the range of } f,$$

$$f^{-1}(f(x)) = x \text{ for all } x \text{ in the domain of } f.$$

Example 1 $y = f(x) = 2x + 3$ is defined for all values of x. The inverse function $x = f^{-1}(y) = \frac{1}{2}(y - 3)$.

Example 2 $f(x) = x^2$ is defined for all values of x, but is not one-to-one unless x is restricted to positive values. Then the inverse function is $\sqrt{x}$, $x \geqslant 0$.

Example 3 The functions $\sin x$ and $\sin^{-1} x$ are inverses for all x.

Example 4 e^x and $\log x$ are inverse functions for $x > 0$.

Theorem If f is continuous and strictly increasing in a closed interval $[a, b]$, then the inverse function is continuous and strictly increasing in the range of f.

Proof. Since f is strictly increasing it is one-to-one and so possesses an inverse. Let $y = f(x), f(a) = \alpha$ and $f(b) = \beta$, then since f is strictly increasing $\alpha < \beta$. By the intermediate value theorem, f takes at least once each value between α and β for $a < x < b$. So that if $y_1 = f(x_1), y_2 = f(x_2)$ and $y_1 < y_2$, then $x_1 \neq x_2$.

If $x_1 > x_2$, then $f(x_1) > f(x_2)$, since f is strictly increasing; this contradicts the hypothesis $y_1 < y_2$. Therefore $x_1 < x_2$ and $f^{-1}(y)$ is strictly increasing.

To prove the continuity, let γ be any number between α and β, that is $\alpha < \gamma < \beta$; then by the intermediate value theorem, $f(c) = \gamma$ for at least one c such that $a < c < b$. Now for any $\epsilon > 0$, let $f(c - \epsilon) = y_1$ and $f(c + \epsilon) = y_2$. Then since f is increasing, $y_1 < \gamma < y_2$, and if y is any number between y_1 and y_2,

$$f^{-1}(y_1) < f^{-1}(y) < f^{-1}(y_2),$$

since f^{-1} is also increasing. Hence

$$|f^{-1}(y) - f^{-1}(\gamma)| = |x - c| < \epsilon,$$

$$\text{for } |y - \gamma| < \delta = \min(\gamma - y_1, y_2 - \gamma).$$

A similar argument will establish one-sided continuity at α and β; this is left as an exercise.

3.7 Uniform continuity

A function f is continuous at $x = a$ if

$$\forall \epsilon > 0; \exists \delta > 0. \; |f(x) - f(a)| < \epsilon \qquad \text{for all } x \text{ in } |x - a| < \delta.$$

The value of δ depends on the choice of ϵ and also, in general, on the point a, Thus, in Examples 1, Section 3.1 we found

$$\delta = \min\{1, \tfrac{1}{3}\epsilon/(1 + |a| + |a|^2)\}.$$

When it is necessary to emphasise this relation we write $\delta = \delta(\epsilon, a)$.

We have said that a function is continuous in an interval if the function is continuous at each point of the interval. Consequently we may expect a different δ for each point of the interval. If, however, it is possible to find a single δ which will suffice for each point of the interval, then we say that the continuity is *uniform* in that interval.

The function f is *uniformly continuous* in an interval I if

$$\forall \epsilon > 0; \exists \delta > 0.\ |f(x) - f(x')| < \epsilon$$

for any pair of points

$$x, x' \in I \quad \text{such that} \quad |x - x'| < \delta.$$

In other words $\delta = \delta(\epsilon)$ is independent of the choice of point x' of the interval.

Example 1 $f(x) = x^3$ is uniformly continuous in $0 \leqslant x \leqslant 1$, since

$$\begin{aligned} |f(x) - f(x')| &= |x^3 - x'^3| \\ &= |x - x'|\,|x^2 + xx' + x'^2| \\ &< 3|x - x'|, \\ &< \epsilon \qquad \text{if } |x - x'| < \delta = \tfrac{1}{3}\epsilon. \end{aligned}$$

Since δ is independent of the choice of x or x', the continuity is uniform in $[0, 1]$.

Example 2 $f(x) = 1/x$ is not uniformly continuous in $0 < x \leqslant 1$, since

$$\begin{aligned} |f(x) - f(x')| &= |x - x'|/xx' \\ &< \epsilon \qquad \text{for } |x - x'| < \delta \end{aligned}$$

if $\delta < \epsilon x'(x' - \delta)$; that is if $\delta < \epsilon x'^2/(1 + \epsilon x')$. Now if x' is taken near to the origin δ is small; indeed $\inf\limits_{0 < x \leqslant 1} \delta = 0$. Therefore f cannot be uniformly continuous in this interval.

Example 3 $f(x) = 1/x$ is uniformly continuous in $0 < \alpha \leqslant x \leqslant 1$. Using the result in the previous example we see that

$$\inf_{\alpha \leqslant x \leqslant 1} \delta = \frac{\epsilon\alpha^2}{1 + \epsilon\alpha},$$

and this δ may be used for any x, x' in the interval.

From the definition it is clear that when a function is uniformly continuous in an interval it is continuous there. The converse is also true when the interval is closed. We shall prove this theorem in Section 5.10.

EXERCISES 3(c)

In Nos. 1–5 find the inverse function f^{-1}:

1. $f(x) = mx + c, m \neq 0$
2. $f(x) = \sqrt{x}, x \geqslant 0$
3. $f(x) = 1/x, x > 0$
4. $f(x) = (x+1)/(x-2), x > 2$
5. $f(x) = x/(1+x^2), |x| \leqslant 1.$

In Nos. 6–9 show that the given function is uniformly continuous in the given interval.

6. $f(x) = x^2, 0 \leqslant x \leqslant 2$
7. $f(x) = \sqrt{x}, 1 \leqslant x \leqslant 2$
8. $f(x) = \sqrt{x}, 0 \leqslant x \leqslant 1$
9. $f(x) = 1/x, x \geqslant 1.$

In Nos. 10–11 show that the given function is not uniformly continuous in the given interval.

10. $f(x) = x^2, x \geqslant 1$
11. $f(x) = \sin(1/x), 0 < x < 1.$

CHAPTER 4

Derivatives

4.1 Definition of derivative

Let f be a function defined in a neighbourhood of a point a. If the limit

$$\lim_{h \to 0} \frac{f(a+h)-f(a)}{h}$$

exists, then f is said to be *differentiable* at $x = a$ and the value of the limit is the *derivative* of the function at $x = a$. We write this limit $f'(a)$.

An equivalent form of the limit which is sometimes more convenient to use is

$$\lim_{x \to a} \frac{f(x)-f(a)}{x-a} = f'(a),$$

whenever the limit exists. Geometrically, $f'(a)$ represents the gradient of the curve $y = f(x)$ at the point where $x = a$.

Other notations used to denote the derivative of $y = f(x)$ at any point x are y', Df or Dy and dy/dx.

Example 1 If f is constant, then $f' = 0$ for all x.

For $f(x) - f(a) = 0$ for all x and a.

Example 2 If $f(x) = x^3$, then $f'(x) = 3x^2$ for any real x. For any $x = a$,

$$\frac{(a+h)^3 - a^3}{h} = 3a^2 + 3ah + h^2$$

$$\to 3a^2 \text{ as } h \to 0.$$

Example 3 If $f(x) = x^{1/2}$, $x \geqslant 0$, then $f'(x) = \frac{1}{2}x^{-1/2}$ if $x > 0$.

For any $a > 0$,

$$\frac{\sqrt{x} - \sqrt{a}}{x - a} = \frac{1}{\sqrt{x} + \sqrt{a}} \to \frac{1}{2\sqrt{a}} \qquad \text{as } x \to a.$$

Example 4 The previous examples are particular cases of the derivative of x^n. Using Exercises 2(e), No. 5 we see immediately that the derivative of x^n is nx^{n-1}, when n is a rational number.

Example 5 $f(x) = |x|$ has no derivative at $x = 0$, for

$$\lim_{x \to 0} |x|/x \text{ does not exist.}$$

However,

$$f'(x) = \begin{cases} +1 \text{ when } x > 0, \text{ and} \\ -1 \text{ when } x < 0. \end{cases}$$

Example 6 The function

$$f(x) = \begin{cases} x \text{ when } x \text{ is irrational,} \\ 0 \text{ when } x \text{ is rational} \end{cases}$$

does not possess a derivative anywhere.

Notice that continuity is a necessary but not sufficient condition for the existence of the derivative of a function at a point. That is, if $f'(a)$ exists then f is continuous at $x = a$. For

$$\lim_{h \to 0} \frac{f(a+h) - f(a)}{h}$$

can exist only if $f(a+h) \to f(a)$ as $h \to 0$, which implies f is continuous at a. The converse is false; the function $|x|$ is continuous at $x = 0$ but is not differentiable there.

4.2 One-sided derivatives

If, in the definition of $f'(a)$, we restrict h to positive or to negative values only, we obtain the one-sided derivatives at a. The *left-hand derivative* of f at a, $f'_-(a)$, is defined to be

$$\lim_{h \to 0-} \frac{f(a+h) - f(a)}{h}$$

when this limit exists.

Similarly

$$f'_+(a) = \lim_{h \to 0+} \frac{f(a+h) - f(a)}{h}$$

whenever this limit exists.

Example When $f(x) = |x|$, $f'_-(0) = -1$ and $f'_+(0) = +1$.

Clearly, for the existence of $f'(a)$ it is necessary and sufficient for $f'_-(a) = f'_+(a)$.

EXERCISES 4(a)

In Nos. 1–6 use the definition to find derivatives of the given functions at any point x:

1. x^4 2. $x^{-1/2}$ 3. $x^{3/2}$

4. $\sin 2x$ 5. $\tan x$ 6. $\operatorname{ch} x$.

7. Show that $f'(0)$ does not exist when $f(x) = x^{1/3}$.

8. Show that $f(x) = x \sin(1/x), x \neq 0, f(0) = 0$ is continuous at $x = 0$ but $f'(0)$ does not exist.

9. Find $f'(0)$ when $f(x) = x^2 \sin(1/x), x \neq 0, f(0) = 0$.

In Nos. 10–12 find the left- and right-hand derivatives of the given functions at the stated point.

10. $f(x) = x \tan^{-1}(1/x), x \neq 0, f(0) = 0$ at $x = 0$.

11. $g(x) = x - [x]$, at $x = 2$.

12. $h(x) = x/(1 + e^{1/x}), x \neq 0, h(0) = 0$ at $x = 0$.

4.3 Rules for differentiation

Theorem Suppose the functions f and g are both differentiable at some point x. Then, at that point,

(a) $f + g$ has the derivative $f' + g'$;

(b) fg has the derivative $f'g + fg'$;

(c) f/g has the derivative $(f'g - fg')/g^2$,

provided $g(x) \neq 0$.

Proof. The proof of (a) is left to the reader. Let $\phi(x) = f(x)g(x)$. Then

$$\frac{\phi(x+h)-\phi(x)}{h} = \frac{f(x+h)\,g(x+h)-f(x)\,g(x)}{h}$$

$$= \frac{f(x+h)-f(x)}{h}\,g(x+h)+f(x)\,\frac{g(x+h)-g(x)}{h}$$

$$\to f'(x)\,g(x)+f(x)\,g'(x) \text{ as } h \to 0.$$

In the final step we have used the limit theorem (b) of Section 2.8, and also the fact that when a function possesses a derivative at a point it must be continuous at that point.

In place of (c) we will prove that the derivative of $1/g, g(x) \neq 0$ is $-g'/g^2$. The derivative of the quotient follows by applying (b) to the product $f(1/g)$.

$$\left(\frac{1}{g(x+h)}-\frac{1}{g(x)}\right)\Big/ h = -\frac{g(x+h)-g(x)}{h}\,\frac{1}{g(x)\,g(x+h)}$$

$$\to -\frac{g'(x)}{[g(x)]^2} \qquad \text{as } h \to 0,$$

again using the limit theorem of Section 2.8.

Another useful rule is that for the derivative of a *function of a function*. If $u = g(x)$ possesses a derivative at a point $x = x_0$, and $y = f(u)$ has a derivative at $u_0 = g(x_0)$, then $y' = f'(u_0)\,g'(x_0)$. Now

$$\frac{f(g(x))-f(g(x_0))}{x-x_0} = \frac{f(u)-f(u_0)}{x-x_0}$$

$$= \frac{f(u)-f(u_0)}{u-u_0}\,\frac{g(x)-g(x_0)}{x-x_0},$$

since $u = g(x)$.

We cannot yet proceed to the limit as $x \to x_0$ for $u - u_0 \equiv g(x) - g(x_0)$ may vanish at points in the neighbourhood of x_0. However, we know that f is a continuous function of u at u_0 and that $f'(u_0)$ exists. So we can write

$$\frac{f(u)-f(u_0)}{u-u_0} = f'(u_0)+\epsilon, \qquad \text{where } \epsilon \to 0 \text{ as } u \to u_0.$$

Then
$$\frac{f(g(x))-f(g(x_0))}{x-x_0} = (f'(u_0)+\epsilon)\,\frac{g(x)-g(x_0)}{x-x_0}$$

$$\to f'(u_0)\,g'(x_0) \qquad \text{as } x \to x_0.$$

The result looks obvious when written in the 'd' notation:

$$\frac{dy}{dx} = \frac{dy}{du}\frac{du}{dx}.$$

It must be remembered, however, that we have not defined dy/dx as a quotient and cannot, therefore, do any cancelling.*

We assume that the reader is familiar with the derivatives of the elementary functions of analysis shown in Table 3.1.†

Table 3.1

f	f'	f	f'
x^n	nx^{n-1}	$\sin x$	$\cos x$
e^x	e^x	$\cos x$	$-\sin x$
$\log x$	$1/x$	$\tan x$	$\sec^2 x$
$\text{sh}\, x$	$\text{ch}\, x$	$\text{ch}\, x$	$\text{sh}\, x$

4.4 Second and higher derivatives

If the derivative of a function f exists at every point of some neighbourhood of a point a, then it may be possible to form the derivative of the derived function. If this is the case it is called the *second derivative* of f and is written f''.

Thus

$$f''(a) = \lim_{h \to 0} \frac{f'(a+h) - f'(a)}{h},$$

provided this limit exists.

If the derivative of the second derivative exists it is called the third derivative of f and written f''' or $f^{(3)}$. This process may be continued to yield 4th, 5th, . . . , nth derivatives, $f^{(4)}, f^{(5)}, \ldots, f^{(n)}$ for any positive integer n.

Other notations for nth derivatives are $y^{(n)}, D^n y, d^n y/dx^n$.

Example 1 If $f(x) = x^m$, then

$$f^{(n)}(x) = m(m-1)(m-2)\cdots(m-n+1)x^{m-n}.$$

When m is a positive integer $f^{(m)}(x) = m!$, and $f^{(n)}(x) = 0$ if $n > m$.

* In Section 10.9 we define the differentials dx and dy and obtain $y' = dy \div dx$.

† See, for example, M.D. Hatton, *Elementary Mathematics for Scientists and Engineers*, Pergamon, 1965.

Example 2 $$D^n\left(\frac{2a}{x^2-a^2}\right) = D^n\left(\frac{1}{x-a}-\frac{1}{x+a}\right)$$
$$= (-1)^n\, n!\,[(x-a)^{-n-1}-(x+a)^{-n-1}].$$

Example 3 $D \sin x = \cos x = \sin(x+\frac{1}{2}\pi)$.

By induction we have

$$D^n \sin x = \sin(x+\tfrac{1}{2}n\pi) \qquad \text{for any positive integer } n.$$

Similarly

$$D^n \cos x = \cos(x+\tfrac{1}{2}n\pi).$$

A formula that is sometimes useful when finding the nth derivative of a product may be written

$$D^n uv = vD^n u + \binom{n}{1} DvD^{n-1}\,u + \binom{n}{2} D^2 vD^{n-2}\,u + \cdots$$
$$+\binom{n}{r} D^r vD^{n-r}\,u + \cdots + \binom{n}{n-1} D^{n-1}\,vDu$$
$$+(D^n v)u$$
$$= \sum_{r=0}^{n} \binom{n}{r} D^r vD^{n-r}u,$$

and is known as *Leibniz's formula*. Here we assume that u and v are functions of x possessing derivatives of order n and we write $D^0 u = u$.

Also

$$\binom{n}{r} = \frac{n(n-1)(n-2)\cdots(n-r+1)}{r!} = \frac{n!}{(n-r)!r!}$$

since r and n are both positive integers;

$$\binom{n}{0} = 1 = \binom{n}{n}.$$

That is $\binom{n}{r}$ is the coefficient of x^r in the binomial expansion of $(1+x)^n$.

Leibniz's formula may be proved by induction and requires the algebraic identity

$$\binom{n}{r}+\binom{n}{r-1} = \binom{n+1}{r}.$$

Example 1 $D^3(x^3e^x) = x^3e^x + \binom{3}{1}3x^2e^x + \binom{3}{2}6xe^x + \binom{3}{3}6e^x$
$$= (x^3+9x^2+18x+6)e^x.$$

Example 2 $D^6(x^2 \sin x) = x^2 D^6 \sin x + \binom{6}{1} Dx^2 D^5 \sin x$

$$+ \binom{6}{2} D^2 x^2 D^4 \sin x,$$

$$(\text{since } D^n x^2 = 0, n = 3, 4, 5, 6)$$

$$= (30 - x^2) \sin x + 12x \cos x.$$

Example 3 $D^n(x^3 e^x) = e^x \{x^3 + \binom{n}{1} Dx^3 + \binom{n}{2} D^2 x^3 + \binom{n}{3} D^3 x^3\},$

$$(\text{since } D^m x^3 = 0, m \geqslant 4)$$

$$= [x^3 + 3nx^2 + 3n(n-1)x + n(n-1)(n-2)]\, e^x.$$

Example 4 The nth derivative of the equation $(1-x^2)y'' - xy' = 0$ may be found by using Leibniz's formula on each term;

$$(1-x^2)y^{(n+2)} + \binom{n}{1}(-2x)y^{(n+1)} + \binom{n}{2}(-2)y^{(n)} - xy^{(n+1)} - \binom{n}{1}.1.y^{(n)} = 0.$$

That is $(1-x^2)y^{(n+2)} - (2n+1)xy^{(n+1)} - n^2 y^{(n)} = 0.$

EXERCISES 4(b)

1. Find the derivatives of:

(a) $\left(\dfrac{x-1}{x+1}\right)^{1/2}$ (b) $(x^2-1)^2(2-x^2)^4$ (c) $x^2\sqrt{(x+1/x)}$

(d) $e^{-3x} \sin 2x$ (e) $\exp(\tan \frac{1}{3}x)$ (f) $\dfrac{1+\sin x}{1-\sin x}$

(g) $\log \sin x$ (h) $x \sin(\log x)$ (i) $\log \sqrt{\left(\dfrac{x-1}{x+1}\right)}$

(j) $\tan^{-1}(x^2)$ (k) $x^2 \sin^{-1}(2x)$ (l) $\operatorname{cosec}^{-1} x$

(m) $\tan^{-1}(\sec x + \tan x)$ (n) $x^{\sin x}$

(o) $\sin^{-1}\left\{\dfrac{x^2}{\sqrt{(x^4+a^4)}}\right\}.$

2. Find the derivatives of:

(a) $x \operatorname{ch} x$ (b) $x^2 \operatorname{sh} x$ (c) $x^{1/2} \operatorname{th} \frac{1}{2}x$

(d) $\log(\operatorname{sech} x - \operatorname{th} x)$ (e) $e^{-2x} \operatorname{sh} 3x$

(f) $\operatorname{ch}^{-1}\sqrt{x}$ (g) $x \operatorname{th}^{-1} x + \log\sqrt{(1-x^2)}$

(h) $2x\sqrt{(x^2+1)}\operatorname{sh}^{-1} x$ (i) $\operatorname{sech}^{-1}(e^{-x})$

(j) $\text{th}^{-1}(\tan \frac{1}{2}x)$ (k) $\text{cosech}^{-1}(\cot x)$

(l) $\text{sh}^{-1}\left\{\tan\left(\frac{1+x}{1-x}\right)\right\}$

3. If $y = \sec x$, prove that

$$y\frac{d^2y}{dx^2} = y^4 + \left(\frac{dy}{dx}\right)^2.$$

4. If $x = x(t), y = y(t)$ show that

$$\frac{dy}{dx} = \frac{\dot{y}}{\dot{x}} \quad \text{and} \quad \frac{d^2y}{dx^2} = \frac{\dot{x}\ddot{y} - \ddot{x}\dot{y}}{\dot{x}^3},$$

where dots denote derivatives with respect to t.

5. Find d^2y/dx^2 when $x = a(1 - \cos t), y = a \sin t$.

6. Use Leibniz's theorem to find the following derivatives:

(a) $D^4(x^2e^{-x})$ (b) $D^5(x^3e^{2x})$

(c) $D^4(x^2 \cos x)$ (d) $D^6(x^3 \log x)$

(e) $D^3(e^{-2x}\tan^{-1} x)$ (f) $D^n(x^2e^{-x}), n > 2$

(g) $D^n(x^3e^{2x}), n > 3$ (h) $D^n(x^2 \cos x), n > 2.$

7. Show that, at $x = 0$, the nth derivative of $x^3/(x^2 - 1)$ is zero if n is even, and is $-n!$ if n is odd and greater than 1.

8. Show that

$$D^{n+1}xy = (n+1)D^ny + xD^{n+1}y.$$

Prove that

$$D^n(x^{n-1}e^{1/x}) = (-1)^nx^{-n-1}e^{1/x}.$$

9. Show that

$$D^n(e^{ax}\cos bx) = r^ne^{ax}\cos(bx + n\alpha),$$

where $r^2 = a^2 + b^2$ and $\alpha = \tan^{-1}(b/a)$.

10. Given $y = \tan^{-1} x$, show that $(1 + x^2)y'' + 2xy' = 0$. Hence determine the value of $y^{(n)}(0)$.

11. Show that $y = \cos(m \sin^{-1} x)$ satisfies the equation

$$(1 - x^2)y'' - xy' + m^2y = 0.$$

Use Leibniz's theorem to differentiate this equation n times. Deduce the value of $y^{(n)}(0)$.

4.5 Sign of $f'(a)$

Theorem If $f'(a) > 0$, then f is strictly increasing in a neighbourhood of $x = a$.

Proof. We have to show that there is a $\delta > 0$ such that

$$f(x) < f(a) \qquad \text{for } a - \delta < x < a,$$

and

$$f(x) > f(a) \qquad \text{for } a < x < a + \delta.$$

Since $f'(a)$ exists,

$$\frac{f(a+h) - f(a)}{h} = f'(a) + \epsilon, \qquad \text{where } \epsilon \to 0 \text{ as } h \to 0.$$

Also $f'(a) > 0$, consequently there is a $\delta > 0$ such that

$$f'(a) + \epsilon > 0 \qquad \text{for } |h| < \delta.$$

That is

$$\frac{f(a+h) - f(a)}{h} > 0 \qquad \text{for } |h| < \delta,$$

and taking h first positive and then negative leads to the required inequalities.

Similarly, if $f'(a) < 0$ then

$$f(x) > f(a) \qquad \text{for } a - \delta < x < a$$

and

$$f(x) < f(a) \qquad \text{for } a < x < a + \delta,$$

for some $\delta > 0$. That is, f is *strictly decreasing* in some neighbourhood of a.

Note that the theorem is true when $f'(a)$ is infinite provided f' has a definite sign.

An immediate deduction from this theorem is that if

$$f(x) \leqslant f(a) \qquad \text{for } |x - a| < \delta$$

and $f'(a)$ exists, then $f'(a) = 0$.

4.6 Extreme values

A point where $f'(x) = 0$ is called a *stationary value* of f. If $f(x) - f(a)$ is of constant sign for all x in a neighbourhood of a, then f has an *extreme value* at $x = a$.

When

$$f(x)-f(a)<0 \qquad \text{for } 0<|x-a|<\delta$$

the extreme value is a *maximum*; if the sign is reversed the extreme value is a *minimum*.

Theorem If f has an extreme value at $x=a$ and $f'(a)$ exists, then $f'(a)=0$.

Proof. If f has an extreme value at $x=a$, $f(x)-f(a)$ is of constant sign for $0<|x-a|<\delta$; let the sign be positive. Then for $h>0$,

$$\frac{f(a+h)-f(a)}{h}>0$$

and

$$\lim_{h\to 0}\frac{f(a+h)-f(a)}{h}\geqslant 0$$

if the limit exists.

Similarly

$$\frac{f(a-h)-f(a)}{-h}<0 \quad \text{and} \quad \lim_{h\to 0}\frac{f(a-h)-f(a)}{-h}\leqslant 0,$$

if it exists.

If $f'(a)$ exists, both these limits must yield the same value; the only possible value is therefore zero.

It should be noted that this condition is necessary but not sufficient for the existence of an extreme value. Thus, if $f(x)=x^3$, $f'(0)=0$ but the origin is *not* an extreme value, for $f(x)-f(0)=x^3$ changes sign with x.

Note also that $f'(a)$ need not exist at an extreme value. For example, both $f(x)=|x|$ and $g(x)=x^{2/3}$ have minima at the origin.

Theorem Sufficient conditions for a minimum when $f'(a)$ exists are $f'(a)=0$ and $f'(a-h)<0$, $f'(a+h)>0$ for $0<h<\delta$.

For a maximum reverse the inequalities.

Example If $f(x)=x^2(x+1)^3$, then

$$\begin{aligned} f'(x) &= 5x(x+\tfrac{2}{5})(x+1)^2 \\ &= 0 \qquad \text{when } x=0, -\tfrac{2}{5}, -1. \end{aligned}$$

For $0<h<\frac{1}{5}$,

$$f'(-h)<0 \quad \text{and} \quad f'(h)>0,$$

consequently f has a minimum at the origin. For $0 < h < \frac{1}{5}$,

$$f'(-\tfrac{2}{5} - h) = (-h)(-2 - 5h)(\tfrac{3}{5} - h)^2 > 0,$$
$$f'(-\tfrac{2}{5} - h) = h(-2 + 5h)(\tfrac{3}{5} + h)^2 < 0.$$

Therefore f has a maximum at $x = -\frac{2}{5}$.

For $0 < h < \frac{3}{5}$,

$$f'(-1 - h) = 5(1 + h)(\tfrac{3}{5} + h)h^2 > 0,$$
$$f'(-1 + h) = 5(1 - h)(\tfrac{3}{5} - h)h^2 > 0.$$

Therefore f has no extreme value at $x = -1$.

EXERCISES 4(c)

Find the extreme values of the functions in Nos. 1–4:

1. $x^4 - 8x^3 + 18x^2 - 14$
2. $16x^4 - 32x^2 + 7$
3. $x^2 + 16/x$
4. $(x^4 + 5x^2 + 8x + 8)e^{-x}$.
5. Show that, in the interval $-3 \leqslant x \leqslant 5$, the least and greatest values of $x^3 - 12x + 20$ are respectively 4 and 85.
6. Show that $y = e^{-ax} \sin bx$ has a sequence of extreme values at values of x which form an arithmetic progression, and that the corresponding values of y form a geometric progression.
7. Show that $(1 + x^2 + ax^4)e^{-x^2}$ has one maximum if $0 \leqslant a \leqslant \frac{1}{2}$. What happens with other values of a?
8. Given that if $f(x) = \alpha x + x^2 \sin(1/x)$, $x \neq 0$ and $f(0) = 0$, show that $f'(0) = \alpha$. Given that $0 < \alpha < 1$, show that there is no interval containing the origin in which f is a steadily increasing function of x.
9. Prove that if f is continuous in a neighbourhood $(a - \delta, a + \delta)$ of a, $f'(x) < 0$ for $(a - \delta, a)$ and $f'(x) > 0$ for $(a, a + \delta)$, then f has a minimum at $x = a$.
10. Show that if $f(x) = 2 + x^{2/3}$ and $g(x) = \exp(-1/x^2)$, $x \neq 0$, $g(0) = 0$, both f and g have minima at the origin.

4.7 Rolle's theorem

Theorem If f is continuous in the closed interval $[a, b]$, f' exists in the open interval (a, b) and $f(a) = f(b)$, then there is a point ξ, $a < \xi < h$, at which $f'(\xi) = 0$.

Proof. We can without loss in generality take $f(a) = 0 = f(b)$, for if $f(a) = f(b) = k$ we can consider the function $g(x) = f(x) - k$. If $f \equiv 0$, then $f' = 0$ for every ξ of (a, b).

If $f \not\equiv 0$, then since f is continuous in $[a, b]$, f attains its supremum at some point ξ of (a, b) (see Section 3.5). By the theorem of Section 4.5, $f'(\xi)$ cannot be positive or negative and so $f'(\xi) = 0$.

Geometrically Rolle's theorem means that for a function satisfying the conditions of the theorem there is at least one point of the interval (there may be more than one such point) at which the tangent to the curve is parallel to the x-axis).

We can use this theorem to extend the theorem of Section 4.5 from the local behaviour of f in the neighbourhood of a point a to the behaviour of f in an interval.

Theorem If f is continuous in the closed interval $[a, b]$ and $f'(x) > 0$ in the open interval (a, b), then f is strictly increasing throughout the closed interval $[a, b]$.

Proof. Let x_1, x_2 be any two points of the interval such that $a < x_1 < x_2 < b$. Then $f(x_1) \neq f(x_2)$, for otherwise $f'(\xi) = 0$ for $x_1 < \xi < x_2$ by Rolle's theorem.

Suppose $f(x_1) > f(x_2)$. Since $f'(x_1) > 0$, there is a point x_3, $x_1 < x_3 < x_2$ such that $f(x_3) > f(x_1)$, by the theorem of Section 4.5. Since f is continuous, the intermediate-value theorem shows that there is a point x_4, $x_3 < x_4 < x_2$, such that $f(x_4) = f(x_1)$. But then, by Rolle's theorem $f'(\xi) = 0$ for $x_1 < \xi < x_4$. This contradicts the hypothesis $f'(x) > 0$ throughout (a, b). Therefore $f(x_1) < f(x_2)$.

Finally we examine the case when $x_1 = a$ and $a < x_2 < b$. Then $f(a) = \lim_{x \to a+} f(x) \leqslant f(x_2)$, since $f(x_1) < f(x_2)$ in (a, b). But if $f(a) = f(x_2)$, then by Rolle's theorem $f'(\xi) = 0$, $a < \xi < x_2$ which is not possible. This proves the theorem.

Example 1 $\sin x < x$ and $\cos x > 1 - \frac{1}{2}x^2$ when $x > 0$.

Let $f(x) = x - \sin x$, then $f'(x) = 1 - \cos x \geqslant 0$ for all x. Hence f is an increasing function of x. Also $f(0) = 0$, therefore $f(x) > 0$ for $x > 0$. That is, $\sin x < x$ for $x > 0$.

Now let $g(x) = \cos x - 1 + \frac{1}{2}x^2$. Then $g'(x) = f(x)$. We have shown that $f(x) > 0$ for $x > 0$, and so g is an increasing function for $x > 0$. Since $g(0) = 0$, therefore $g(x) > 0$ for $x > 0$. That is $\cos x > 1 - \frac{1}{2}x^2$ for $x > 0$.

Example 2 $f(x) = (2x^2 + 4x + 3)e^{-2x}$ is a decreasing function for all x. For $f'(x) = -2(2x^2 + 2x + 1)e^{-2x} = -4e^{-2x}\{(x + \frac{1}{2})^2 + \frac{1}{4}\}$. Clearly $f'(x) < 0$ for all x, and so f is a decreasing function of x for all x.

4.8 The mean value theorem

Theorem If f is continuous in $[a, b]$ and f' exists in (a, b), then there is a point ξ, $a < \xi < b$, for which

$$f(b) - f(a) = (b - a)f'(\xi).$$

Proof. This follows directly by applying Rolle's theorem to the auxiliary function

$$\phi(x) = f(a) - f(x) + \left(\frac{x-a}{b-a}\right)[f(b) - f(a)].$$

Geometrically, we notice that ϕ represents the difference between the ordinates to the curve $y = f(x)$ and to the chord joining the ends of the curve. The theorem states that there is at least one point at which the tangent is parallel to this chord.

The result of the mean value theorem can also be written in the form

$$f(a + h) - f(a) = hf'(a + \theta h), \qquad 0 < \theta < 1.$$

Example 1 If $f'(x) = 0$ for $a < x < b$, then f is a constant in $[a, b]$.

The existence of f' ensures the continuity of f, so that f satisfies the conditions of the mean value theorem. Let x_1, x_2 be any two points such that $a \leqslant x_1 < x_2 \leqslant b$. Then we have

$$f(x_2) - f(x_1) = (x_2 - x_1)f'(\xi), \qquad x_1 < \xi < x_2.$$

But $f'(\xi) = 0$, and so $f(x_1) = f(x_2)$. That is f is constant throughout $[a, b]$.

Example 2 If $f'(x) > 0$ for $a < x < b$, then f is strictly increasing in $[a, b]$.

With the same choice of x_1, x_2 as before, we have

$$\begin{aligned} f(x_2) - f(x_1) &= (x_2 - x_1)f'(\xi), \qquad x_1 < \xi < x_2 \\ &> 0. \end{aligned}$$

Example 3 If f' exists throughout (a, ∞) and f' tends to a finite limit l as $x \to \infty$, then $f(x)/x \to l$ as $x \to \infty$.

Since $\lim_{x \to \infty} f'(x) = l$, we have for any $\epsilon > 0$,

$$|f'(x) - l| < \tfrac{1}{2}\epsilon \qquad \text{for } x \geqslant \lambda = \lambda(\epsilon).$$

Let $\phi(x) = f(x) - lx$, then ϕ satisfies the conditions for the mean value theorem in the interval $\lambda \leqslant x \leqslant X$, for any $X > \lambda$. Since ϕ' is bounded

$$|\phi(X) - \phi(\lambda)| \leqslant (X - \lambda).\tfrac{1}{2}\epsilon < \tfrac{1}{2}\epsilon X.$$

Dividing by X we have

$$\left|\frac{f(x)}{X} - l - \frac{f(\lambda) - l\lambda}{X}\right| < \tfrac{1}{2}\epsilon.$$

Then
$$\left|\frac{f(X)}{X} - l\right| \leqslant \tfrac{1}{2}\epsilon + \left|\frac{f(\lambda) - l\lambda}{X}\right|$$

$$< \epsilon \qquad \text{if } X > \max\{\lambda, 2|f(\lambda) - \lambda|/\epsilon\}.$$

Therefore
$$\frac{f(X)}{X} \to l \qquad \text{as } X \to \infty.$$

Example 4 If f has a definite limit l as $x \to \infty$, then f' either oscillates or tends to zero as $x \to \infty$.

For if $f(x) \to l$ as $x \to \infty$, then for any $\epsilon > 0$

$$|f(x) - f(X)| < \epsilon \qquad \text{for } x > X.$$

Then, since
$$\frac{f(x) - f(X)}{x - X} = f'(\xi), \qquad X < \xi < x,$$

we have
$$|f'(\xi)| < \frac{\epsilon}{x - X}$$

Therefore, if f' has a limit as $x \to \infty$ it must be zero. However, as $x \to \infty$, ξ may not take *all* values greater than X and so oscillation cannot be excluded.

EXERCISES 4(d)

1. Verify Rolle's theorem when

$$f(x) = (x + 1)^m (x - 1)^n, \qquad -1 \leqslant x \leqslant 1,$$

where m and n are positive integers.

2. Show that the result of Rolle's theorem is not true for

(a) $f(x) = x^{-2}$, $[-1, 1]$, (b) $g(x) = |x|$, $[-1, 1]$.

3. The functions u, v and their derivatives u', v' are continuous in $[a, b]$, and $W(u, v) = uv' - u'v$ never vanishes at any point of $[a, b]$. Show that between any two zeros of u lies one of v and conversely. The determinant

$$W(u, v) = \begin{vmatrix} u & v \\ u' & v' \end{vmatrix}$$

is called the *Wronskian* of u and v.

4. Find values of ξ in $[a, b]$ in the mean value theorem when

$$f(x) = x^k \quad \text{and} \quad k = 1, 2 \text{ and } 3.$$

5. Find $\xi \in (0, \frac{1}{2})$ when the mean value theorem is applied to

$$f(x) = x(x-1)(x-2)$$

in the interval $[0, \frac{1}{2}]$.

6. If f' exists in $a \leqslant x \leqslant b$, $f'(a) = \alpha$, $f'(b) = \beta$ and $\alpha < \gamma < \beta$, show that there is a point $\xi \in (a, b)$ for which $f'(\xi) = \gamma$.

7. Prove that, when $0 < \theta < \frac{1}{2}\pi$,

(a) $\dfrac{2\theta}{\pi} < \sin\theta < \theta$ (b) $1 - \frac{1}{2}\theta^2 < \cos\theta < 1 - \frac{1}{2}\theta^2 + \dfrac{\theta^4}{24}$

8. Show that $f(x) = e^x(x^2 - 6x + 12) - (x^2 + 6x + 12)$ is an increasing function of x in the interval $(0, \infty)$. Prove that

$$0 < \frac{1}{e^x - 1} - \frac{1}{x} + \tfrac{1}{2} < \frac{x}{12} \qquad \text{when } x > 0.$$

9. Show that

$$\frac{\frac{1}{2}x^2}{1+x} < x - \log(1+x) < \tfrac{1}{2}x^2 \qquad \text{for } x > 0,$$

and that the inequalities are reversed if $-1 < x < 0$.

10. Given

$$P_n(x) = 1 + x + \frac{x^2}{2!} + \cdots + \frac{x^n}{n!},$$

prove that $P_n(x) = 0$ has no real root when n is even and exactly one real root when n is odd.

11. Prove that $f_n(x) = D^n(x^2-1)^n$ has n distinct zeros in $(-1, 1)$.

12. Given that f is a function possessing a continuous second derivative and that for all real x and y,

$$f(x) < \tfrac{1}{2}\{f(x+y)+f(x-y)\},$$

prove that $f''(x) > 0$.

13. If f has derivatives of the first two orders, prove that there is a number ξ, $x-h<\xi<x+h$, such that

$$\frac{f(x+h)+f(x-h)-2f(x)}{h^2} = f''(\xi).$$

14. Given that f' satisfies the conditions for Rolle's theorem, show by considering the auxiliary function

$$\phi(x) = f(a)(b-x)(b+x-2a)-f(x)(b-a)^2+f(b)(x-a)^2$$
$$-f'(a)(b-x)(x-a)(b-a),$$

that

$$f(b) = f(a)+(b-a)f'(a)+\tfrac{1}{2}(b-a)^2 f''(\xi), \qquad a<\xi<b.$$

This is known as the *second mean value theorem*

4.9 Cauchy's mean value theorem

Theorem If f and g are continuous in the closed interval $[a, b]$, and f', g' both exist in the open interval (a, b), then there is a point ξ, $a<\xi<b$, for which

$$\frac{f(b)-f(a)}{g(b)-g(a)} = \frac{f'(\xi)}{g'(\xi)},$$

provided g' does not vanish in (a, b).

Proof. Notice that we cannot apply the mean value theorem to each of the functions f and g separately, for this would lead to two different points ξ_1 and ξ_2.

The auxiliary function

$$\phi(x) = \begin{vmatrix} f(x) & g(x) & 1 \\ f(a) & g(a) & 1 \\ f(b) & g(b) & 1 \end{vmatrix},$$

satisfies all the conditions of Rolle's theorem.

Therefore $\phi'(\xi) = 0$ for $a < \xi < b$.

That is

$$\begin{vmatrix} f'(\xi) & g'(\xi) & 0 \\ f(a) & g(a) & 1 \\ f(b) & g(b) & 1 \end{vmatrix} = 0,$$

which yields the required result on simplification.

We can also write $b = a + h$; then the result takes the form

$$\frac{f(a+h) - f(a)}{g(a+h) - g(a)} = \frac{f'(a+\theta h)}{g'(a+\theta h)}, \qquad 0 < \theta < 1.$$

Notice that when $g(x) = x$ this reduces to the mean value theorem.

4.10 Indeterminate forms

In the limit theorem (c) of Section 2.8 we showed that if, as $x \to a$, $f(x) \to \alpha$ and $g(x) \to \beta$, then $f/g \to \alpha/\beta$ provided $\beta \neq 0$. We now examine the case when $\beta = 0 = \alpha$. There are two theorems to consider.

Theorem 1 If $f(a) = g(a) = 0$ and $f'(a), g'(a)$ exist and have a definite ratio, then

$$\lim_{x \to a} \frac{f(x)}{g(x)} = \frac{f'(a)}{g'(a)}.$$

Proof. This follows directly from the definition of a derivative. For

$$\lim_{x \to a} \frac{f(x)}{g(x)} = \lim_{x \to a} \frac{f(x) - f(a)}{g(x) - g(a)}$$

$$= \lim_{x \to a} \left(\frac{f(x) - f(a)}{x - a}\right) \Bigg/ \left(\frac{g(x) - g(a)}{x - a}\right)$$

$$= \frac{f'(a)}{g'(a)}.$$

Theorem 2 If $f'(x), g'(x)$ exist in some neighbourhood of $x = a$, (but not necessarily at a itself) and $g'(x)$ never vanishes in this neighbourhood except possibly at a itself, then

$$\lim_{x \to a} \frac{f(x)}{g(x)} = \lim_{x \to a} \frac{f'(x)}{g'(x)},$$

provided that the limit on the right-hand side of the equation exists.

Proof. Using Cauchy's mean value theorem in the interval $a < x < a + h$ for some $h > 0$, we have

$$\frac{f(x)}{g(x)} = \frac{f(x) - f(a)}{g(x) - g(a)} = \frac{f'(\xi)}{g'(\xi)}, \qquad a < \xi < a + h.$$

Then, provided

$$\lim_{h \to 0+} \frac{f'(\xi)}{g'(\xi)}$$

exists, we have

$$\lim_{x \to a+} \frac{f(x)}{g(x)} = \lim_{x \to a+} \frac{f'(x)}{g'(x)}.$$

Similarly, considering the interval $a - h < x < a$,

$$\lim_{x \to a-} \frac{f(x)}{g(x)} = \lim_{x \to a-} \frac{f'(x)}{g'(x)}$$

provided the limit on the right-hand side exists.

Combining these two results gives the theorem which is known as *l'Hôpital's rule.*

Example 1

$$\lim_{x \to 0} \frac{e^{ax} - e^{-ax}}{\log(1 + x)}.$$

Here $f(x) = e^{ax} - e^{-ax}$ and $g(x) = \log(1 + x)$; $f(0) = g(0)$. Then

$$f'(x) = a(e^{ax} + e^{-ax}), \quad g'(x) = 1/(1 + x)$$

and $f'(0) = 2a$, $g'(0) = 1$. Therefore

$$\lim_{x \to 0} \frac{e^{ax} - e^{-ax}}{\log(1 + x)} = 2a.$$

Example 2

$$\lim_{x \to 0} \frac{x^2}{\operatorname{ch} x - 1}$$

Here

$$f(x) = x^2, f'(x) = 2x, g(x) = \operatorname{ch} x - 1, g'(x) = \operatorname{sh} x.$$

Then

$$\lim_{x \to 0} \frac{x^2}{\operatorname{ch} x - 1} = \lim_{x \to 0} \frac{2x}{\operatorname{sh} x} = 2,$$

using

$$\lim_{x \to 0} (\operatorname{sh} x/x) = 1.$$

Alternatively we could use l'Hôpital's rule a second time to evaluate

$$\lim_{x \to 0} \frac{x}{\operatorname{sh} x} = \frac{1}{\operatorname{ch} 0} = 1.$$

This is an example of a general result that if $f(a) = 0 = g(a)$ and $f'(a) = 0 = g'(a)$, but $\lim_{x \to a} [f''(x)/g''(x)]$ exists, then

$$\lim_{x \to a} \frac{f(x)}{g(x)} = \lim_{x \to a} \frac{f''(x)}{g''(x)}.$$

Indeed we can extend Theorem 2 to give

Theorem 3 If $f^{(r)}(a) = 0 = g^{(r)}(a)$, $r = 0, 1, 2, \ldots, n-1$ and

$$\lim_{x \to a} \frac{f^{(n)}(x)}{g^{(n)}(x)}$$

exists, then

$$\lim_{x \to a} \frac{f(x)}{g(x)} = \lim_{x \to a} \frac{f^{(n)}(x)}{g^{(n)}(x)}.$$

This may be proved by induction; it is left as an exercise.

Example 1

$$\lim_{x \to 0} \frac{e^x - (1+2x)^{1/2}}{\log(1+x^2)}$$

Here

$$f(x) = e^x - (1+2x)^{1/2}, \quad f'(x) = e^x - (1+2x)^{-1/2},$$
$$f''(x) = e^x + (1+2x)^{-3/2};$$

$$g(x) = \log(1+x^2), \quad g'(x) = \frac{2x}{1+x^2}, \quad g''(x) = \frac{2(1-x)^2}{(1+x^2)^2}.$$

Then $f(0) = f'(0) = 0$ and $g(0) = g'(0) = 0$, but $f''(0) = 2$ and $g''(0) = 2$. Therefore

$$\lim_{x \to 0} \frac{e^x - (1+2x)^{1/2}}{\log(1+x^2)} = 1.$$

Example 2 $\lim_{x \to 0} (x^{-1} - \cot x) = \lim_{x \to 0} \dfrac{\sin x - x\cos x}{x \sin x} = 0.$

Here

$$f(x) = \sin x - x\cos x, \quad f(0) = f'(0) = f''(0) = 0,$$

and $g(x) = x\sin x$, $\quad g(0) = g'(0) = 0$ and $g''(0) = 2$.

Note. For the validity of these theorems it is necessary for the limit on the right-hand side of the equation to exist in each case.

Thus if $f(x) = x^2 \sin(1/x), f(0) = 0$ and $g(x) = \tan x$, then $f'(x) = 2x \sin(1/x) - \cos(1/x), x \neq 0, g'(x) = \sec^2 x$. Clearly, $\lim_{x \to 0} f'(x)$ does not exist and so $\lim_{x \to 0} [f'(x)/g'(x)]$ does not exist. But

$$\frac{f(x)}{g(x)} = \frac{x^2 \sin(1/x)}{\tan x} = [x \sin(1/x)] \cdot \frac{x}{\tan x}$$

$$\to 0 \text{ as } x \to 0.$$

EXERCISES 4(e)

1. Verify Cauchy's mean value theorem when $f(x) = x^3, g(x) = x^2$ and $2 \leqslant x \leqslant 4$.

2. Show that the result of Cauchy's mean value theorem is not true for $f(x) = 4x^3 + 6x^2 - 12x, g(x) = 3x^4 + 4x^3 - 6x^2$ in $[0, 1]$.

Evaluate the limits in Nos. 3–17:

3. $\lim_{x \to 0} \frac{x^2}{\operatorname{ch} x - 1}$

4. $\lim_{x \to 0} \frac{(1-x)e^x - 1}{(e^x - 1)^2}$

5. $\lim_{x \to \pi} \frac{1 + \cos x}{\tan^2 x}$

6. $\lim_{x \to 0} \frac{x(2^x - 1)}{1 - \cos x}$

7. $\lim_{x \to 0} \frac{\operatorname{sh} x - x}{\log(1+x) - x + \frac{1}{2}x^2}$

8. $\lim_{x \to 0} (\operatorname{cosec} x - \cot x)$

9. $\lim_{x \to n} (x - n) \operatorname{cosec} \pi x$.

10. $\lim_{x \to \frac{1}{2}\pi} (2x \tan x - \pi \sec x)$

11. $\lim_{x \to 0} (\operatorname{ch} x)^{x^{-2}}$

12. $\lim_{x \to \frac{1}{4}\pi} (\tan x)^{\tan 2x}$

13. $\lim_{x \to 0} \frac{e^x - (1+2x)^{1/2}}{\log(1+x^2)}$

14. $\lim_{x \to 0} \frac{\sin^{-1} x - \sin x}{\tan x - \tan^{-1} x}$

15. $\lim_{\omega \to n} \left(b - \frac{cn\omega}{n^2 - \omega^2} \sin nt + \frac{c\omega^2}{n^2 - \omega^2} \sin \omega t \right)$, b, c, n, t constants

16. $\lim_{n \to \infty} (1 + x/n)^n$

17. $\lim_{n \to \infty} n(x^{1/n} - 1)$.

18. Given that $f(x) = x^m \sin(x^{-n}), x \neq 0, f(0) = 0$, where m and n are positive integers, find when $f'(0)$ exists. For what values of m and n is f' continuous at $x = 0$? Taking $g(x) = x^k$, with a suitable choice of k, construct an example in which f, g are continuous at $x = 0$, $f(0) = 0 = g(0), g'(x) \neq 0$ in $(-a, a)$ and $\lim_{x \to 0} [f(x)/g(x)]$ exists but $\lim_{x \to 0} [f'(x)/g'(x)]$ does not exist.

19. Prove that, if f' exists for $a < x < a + h$, then

$$f(a+h) = f(a) + (e^h - 1)e^{-\theta h} f'(a + \theta h), \qquad 0 < \theta < 1.$$

20. Given that f, g satisfy the conditions of Cauchy's mean value theorem f'', g'' are continuous in (a, b) and $f''g' \neq f'g''$, prove that $\lim_{h \to 0} \theta = \frac{1}{2}$.

4.11 Convex functions

If f is a function defined in $[a, b]$ and such that, for every pair of points x_1, x_2 in $[a, b]$

$$f(\tfrac{1}{2}(x_1 + x_2)) \leqslant \tfrac{1}{2}[f(x_1) + f(x_2)],$$

then f is said to be *convex* in $[a, b]$.

In geometrical terms this means that the curve joining any two points $(x_1, f(x_1))$, $(x_2, f(x_2))$ lies below the chord joining those points.

Theorem If f is continuous in $[a, b]$ and f' exists in the open interval (a, b), then f is convex on (a, b) if and only if f' is increasing in (a, b)

Proof. Let x be any point of (a, b) and h a positive number such that $x \pm h$ lie in $[a, b]$. Then, if f is convex

$$f(x) \leqslant \tfrac{1}{2}[f(x-h) + f(x+h)].$$

This can be rearranged as

$$f(x) - f(x-h) \leqslant f(x+h) - f(x).$$

Then using the mean value theorem we have

$$f'(x - \theta_1 h) \leqslant f'(x + \theta_2 h), \qquad 0 < \theta_1, \theta_2 < 1,$$

and so f' is increasing.

Now, if f' is increasing, we have using the mean value theorem

$$\frac{f(x) - f(x-h)}{h} = f'(x - \theta_1 h), \quad 0 < \theta_1 < 1$$

$$\frac{f(x+h) - f(x)}{h} = f'(x + \theta_2 h), \quad 0 < \theta_2 < 1.$$

Increasing f' implies that

$$f'(x-\theta_1 h) \leqslant f'(x+\theta_2 h),$$

and this implies that

$$f(x)-f(x-h) \leqslant f(x+h)-f(x),$$

or

$$f(x) \leqslant \tfrac{1}{2}[f(x+h)+f(x-h)].$$

That is f is convex.

From this result we can deduce the following.

Corollary If f'' exists in (a, b) then f is convex on $[a, b]$ if and only if $f''(x) \geqslant 0$ in (a, b).

4.12 Point of inflexion

A point of a graph where the curve crosses its tangent at this point is called a *point of inflexion*. This means that the direction of convexity of the curve changes at such a point. Consequently f'' must change sign as x increases through the point; at the point of inflexion f'' vanishes. Thus in the example of Section 4.6, $f(x) = x^2(x+1)^3$ has a point of inflexion at $x = -1$. For

$$\begin{aligned} f''(x) &= 2(x+1)(10x^2+8x+1); \\ f''(-1) &= 0, \\ f''(-1-h) &= -20h\,[(\tfrac{3}{5}+h)^2 - \tfrac{3}{50}] < 0, \\ f''(-1+h) &= 20h\,[(\tfrac{3}{5}-h)^2 - \tfrac{3}{50}] > 0. \end{aligned}$$

The second derivative may be used to write down sufficient conditions for an extreme value. Thus f has a *minimum at* $x = a$ *if*

$$f'(a) = 0 \quad \text{and} \quad f''(a) > 0.$$

For a maximum the inequality must be reversed.

Example If $f(x) = x^2(x+1)^3$, then $f'(0) = 0$ and $f''(0) > 0$; therefore the function has a minimum at the origin.

EXERCISES 4(f)

1. Show that if f has a continuous positive second derivative in an interval (a, b), then f is convex in that interval.

2. Show that if f is convex in $[a, b]$ and possesses a second derivative, then $f'' \geqslant 0$ in (a, b).

3. Another definition of a convex function is a function for which $(x_2 - x_1)f(x) < (x_2 - x)f(x_1) + (x - x_1)f(x_2)$, for points $x_1 < x < x_2$ of (a, b). Prove that this definition implies the continuity of f in (a, b).

CHAPTER 5

Sequences

5.1 Definition of a sequence

In Chapter 2 we considered functions whose domain is the set of real numbers; we now restrict the domain to the set of positive integers. These functions of a positive integral variable n are usually called *sequences*, and are often denoted by (x_n) or $\{x_n\}$ to emphasise that to each of the positive integers $1, 2, 3, \ldots, n, \ldots$ there corresponds one of the numbers $x_1, x_2, x_3, \ldots, x_n, \ldots$. If the terms of the sequence are also functions of x we sometimes write the sequence $\{f_n(x)\}$.

Examples

1. $\frac{1}{2}(1 - 1/n)$ 2. a^n
3. $\frac{1}{2}[1 - (-1)^n]$ 4. $\sin \frac{1}{4}n\pi$
5. $1 + (-\frac{1}{2})^n$ 6. $n/(n^2 - 2)$
7. $1{\cdot}4, 1{\cdot}41, 1{\cdot}414, 1{\cdot}4142, \ldots$ that is x_n is the decimal approximation to $\sqrt{2}$ terminating with the nth decimal.
8. $(-1)^n (1 + 1/n)$ 9. p_n = nth prime number
10. $s_n = 1 + \frac{1}{2} + \frac{1}{3} + \cdots + 1/n$ 11. $[(-1)^n - 1]\, n$
12. $x_n = x_{n-1} + x_{n-2}, n = 3, 4, 5, \ldots, x_1 = x_2 = 1$ (Fibonacci's sequence)

5.2 Bounds of a sequence

If there exists a constant H such that

$$x_n \leqslant H \text{ for all values of } n,$$

then the sequence (x_n) is said to be *bounded above*. Similarly the sequence is *bounded below* if there is a constant h such that

$$x_n \geqslant h \text{ for all values of } n.$$

A sequence which is bounded above and below is said to be *bounded*; in this case we can write

$$|x_n| \leqslant K \text{ for all } n,$$

where $K = \max(|h|, |H|)$.

In the above examples Nos. 1, 3, 4, 5, 6, 7, 8 are bounded; Nos. 9, 10 and 12 are bounded below (all are positive) but not bounded above; No. 11 is bounded above but not below.

We can define the exact bounds of a bounded sequence as we did for a bounded function in Section. 2.4.The *supremum* or *least upper bound* M is such that

$$x_n \leqslant M \text{ for all } n,$$

and

$$\forall \epsilon > 0; x_n > M - \epsilon \qquad \text{for at least one value of } n.$$

We write

$$M = \sup x_n \quad \text{or} \quad \text{lub } x_n.$$

Similarly we have the *infimum* or *greatest lower bound* m satisfying

$$x_n \geqslant m \text{ for all } n,$$

and

$$\forall \epsilon > 0; x_n < m + \epsilon \qquad \text{for at least one value of } n.$$

We write

$$m = \inf x_n \quad \text{or} \quad \text{glb } x_n.$$

5.3 Convergent and divergent sequences

When a sequence possesses a limit as $n \to \infty$ it is said to be a *convergent* sequence. Following the definition of a limit in Chapter 2 we can write:

The sequence (x_n) converges to a limit l if

$$\forall \epsilon > 0; \exists N.\ |x_n - l| < \epsilon \qquad \text{for } n > N.$$

In particular, if $l = 0$, the sequence is said to be *null*.

A sequence (x_n) which does not converge is said to *diverge*; this may be because x_n is unbounded or because x_n does not have a unique limiting value. In the latter case (x_n) is called an *oscillatory* sequence, the oscillation being finite or infinite according as x_n is or is not bounded. Thus $(-1)^n (2 - 1/n)$ oscillates finitely and $(-1)^n n^2$ oscillates infinitely.

Example 1 If $|a| < 1$, then (a^n) and (na^n) are null sequences. Let $1/|a| = 1 + p, p > 0$. Then using Bernoulli's inequality (Section 1.7),

$$\frac{1}{|a|^n} = (1+p)^n > 1 + np > np \qquad \text{for } n > 1.$$

Hence

$$|a|^n < \frac{1}{np} < \epsilon \qquad \text{for } n > N = [1/p\epsilon].^*$$

Again

$$\begin{aligned} n\,|a|^n &= \frac{n}{(1+p)^n} \\ &= \frac{n}{1 + \binom{n}{1}p + \binom{n}{2}p^2 + \cdots + p^n} \\ &< \frac{n}{\binom{n}{2}p^2}, \end{aligned}$$

since all the terms are positive. That is

$$n\,|a|^n < \frac{2}{(n-1)\,p^2} < \epsilon \qquad \text{for } n > 1 + \frac{2}{p^2\epsilon},$$

and so we can take $N = 1 + [2/p^2\epsilon]$.

Example 2 $x_n = n^{1/n} - 1$ is a null sequence.

Now

$$n = (1 + x_n)^n > \binom{n}{2} x_n^2,$$

since all the terms of the expansion are positive. Then

$$x_n^2 < \frac{2}{n-1} < \frac{2}{n - \frac{1}{2}n}, \qquad n > 2,$$

and

$$|x_n| < 2/n^{1/2} < \epsilon \qquad \text{for } n > 4/\epsilon^2,$$

so we can take $N = \max\{2, (4/\epsilon^2)\}$.

Note. This proves that $\lim_{n\to\infty} \sqrt[n]{n} = 1$.

Example 3 $x_n = \dfrac{n+3}{n^2-1}$ is a null sequence.

Clearly $x_n > 0$ for $n > 1$. Then

$$x_n = \frac{n+3}{n^2-1} < \frac{n+n}{n^2-1} \qquad \text{for } n > 3,$$

* The square bracket function $[x]$ is defined in Section 2.1.

$$< \frac{2n}{n^2 - \frac{1}{2}n^2}$$

Therefore

$$x_n < \epsilon \qquad \text{for } n > N = \max\{3, (4/\epsilon)\}.$$

Example 4 $\quad x_n = \dfrac{2n^2+3}{n^2-5}$ converges to 2.

For

$$|x_n - 2| = \frac{13}{|n^2-5|} < \frac{13}{n^2 - \frac{1}{2}n^2} \qquad \text{for } n > 4$$

(more precisely for $n^2 > 10$)

$$< \epsilon \qquad \text{for } n > \sqrt{(26/\epsilon)}.$$

Thus $|x_n - 2| < \epsilon$ for $n > N = \max\{4, [\sqrt{(26/\epsilon)}]\}$.

EXERCISES 5(a)

1. Show that the sequence (x_n) is null when x_n is given by:

(a) $\dfrac{3n+2}{n^2+1}$ (b) $\dfrac{n^2+4}{n^3-12}$ (c) $\dfrac{(-1)^n}{\sqrt{n}}$ (d) $\dfrac{n^3+2n^2-1}{n^4-n^2+2}$.

2. Prove that the following sequences converge:

(a) $\dfrac{n-1}{n+1}$ (b) $\dfrac{2n^2+1}{n^2+3n}$ (c) $\dfrac{3n^2-1}{n^2-5n}$ $(n \geqslant 6)$

(d) $\dfrac{3n^3+3n^2-n-1}{n^3+n^2-2}$ $(n \geqslant 2)$.

3. Find the supremum and infimum of each of the following sequences:

(a) $\dfrac{n-1}{2n}$ (b) $\dfrac{(-1)^n n}{2n+1}$ (c) $\dfrac{1+(-1)^n}{3}$ (d) $\sin \frac{1}{2}n\pi$

(e) $\dfrac{1}{n} - \sin \frac{1}{2}n\pi$ (f) $\left(1 + \dfrac{1}{2n}\right) \cos \frac{1}{3}n\pi$.

4. Show that $(x^{1/n} - 1), x > 0$, is a null sequence. That is, prove that

$$\lim_{n \to \infty} {}^n\sqrt{x} = 1, \text{ when } x > 0.$$

5. Show that if $\lim_{n \to \infty} n(x^{1/n} - 1) = f(x)$, then $f(1) = 0, f(x) \gtrless 0$ according as $x \gtrless 1, f(1/x) = -f(x)$ and $f(xy) = f(x) + f(y)$. [We shall see later that $f(x) = \log x$, see Exercises 4(e) No. 17.]

6. Prove that if (x_n) is null and (a_n) is bounded, then $(a_n x_n)$ is null.

7. Prove that a sequence cannot converge to more than one limit.

5.4 Properties of convergent sequences

Property 1 A convergent sequence is bounded.

Proof. If (x_n) converges to l, then

$$|x_n - l| < \epsilon \quad \text{or} \quad l - \epsilon < x_n < l + \epsilon \qquad \text{for } n > N = N(\epsilon).$$

Choose $\epsilon = \frac{1}{2}|l|$; then

$$\tfrac{1}{2}|l| < |x_n| < \tfrac{3}{2}|l| \qquad \text{for } n > N_1 = N(\tfrac{1}{2}|l|).$$

Then $M = \max \{|x_1|, |x_2|, \ldots, |x_{N_1}|, \frac{3}{2}|l|\}$ is an upper bound.

Property 2 If (x_n) converges to α and (y_n) converges to β, then

(a) $(x_n \pm y_n)$ converges to $\alpha \pm \beta$,

(b) $(x_n y_n)$ converges to $\alpha\beta$,

(c) (x_n/y_n) converges to α/β, provided $\beta \neq 0$.

Compare these results with the limit theorems in Section 2.8. The proofs are very similar and are left as exercises.

Note. It is possible for

$$\lim_{n\to\infty} (x_n + y_n), \qquad \lim_{n\to\infty} x_n y_n, \qquad \lim_{n\to\infty} x_n/y_n$$

to exist even when (x_n) and (y_n) do not converge. For example

$$x_n = (-1)^n, \quad y_n = (-1)^{n-1}(1 + 1/n).$$

Property 3 (Cauchy's theorem on limits). If (x_n) converges to α, then the sequence $y_n = (x_1 + x_2 + \cdots + x_n)/n$ also converges to α.

Proof. Let $x_n = \alpha + s_n$; then (s_n) is a null sequence, and

$$\frac{x_1 + x_2 + \cdots + x_n}{n} = \alpha + \frac{s_1 + s_2 + \cdots + s_n}{n}$$

Hence we must show that

$$\frac{s_1 + s_2 + \cdots + s_n}{n} \to 0 \text{ as } n \to \infty.$$

Since (s_n) is a null sequence,

$$|s_n| < \tfrac{1}{2}\epsilon \text{ for } n > N.$$

Then

$$\left|\frac{s_1 + s_2 + \cdots + s_n}{n}\right| \leqslant \frac{|s_1| + |s_2| + \cdots + |s_N|}{n} + \frac{|s_{N+1}| + |s_{N+2}| + \cdots + |s_n|}{n}, n > N$$

$$< \frac{N\mu}{n} + \frac{(n-N)}{n}\tfrac{1}{2}\epsilon,$$

$$\text{where } \mu = \max_{1 \leqslant r \leqslant N} (|s_r|)$$

$$< \tfrac{1}{2}\epsilon + \tfrac{1}{2}\epsilon \quad \text{for } n > 2N\mu/\epsilon.$$

Therefore

$$\frac{x_1 + x_2 + \cdots + x_n}{n} \to \alpha \qquad \text{as } n \to \infty.$$

Example 1 $(1 + \frac{1}{2} + \cdots + 1/n)/n \to 0$ as $n \to \infty$.

Example 2 If $x_n = \log y_n \to \log \alpha$ as $n \to \infty$, then

$$\frac{x_1 + x_2 + \cdots + x_n}{n} = \frac{1}{n}\log(y_1 y_2 \cdots y_n) \to \log \alpha \text{ as } n \to \infty.$$

Assuming the continuity of the logarithm we have

$$\lim_{n \to \infty} (y_1 y_2 y_3 \cdots y_n)^{1/n} = \alpha,$$

when

$$y_n > 0 \quad \text{and} \quad y_n \to \alpha \text{ as } n \to \infty.$$

Example 3 Since $\dfrac{n}{n-1} \to 1$ as $n \to \infty$, then

$$\sqrt[n]{\left(1 \cdot \frac{2}{1} \cdot \frac{3}{2} \cdots \frac{n}{n-1}\right)} = \sqrt[n]{n} \to 1 \text{ as } n \to \infty.$$

EXERCISES 5(b)

1. Given that (x_n) converges to α and (y_n) converges to β, prove that

   (a) $(x_n \pm y_n)$ converges to $\alpha \pm \beta$, and

   (b) $(x_n y_n)$ converges to $\alpha\beta$.

2. Given that (x_n) converges to α and (y_n) converges to $\beta\ (\neq 0)$, prove that (x_n/y_n) converges to α/β.

3. Given that (x_n) converges to α, (y_n) converges to β and $x_n < y_n$ for all n, prove that $\alpha \leqslant \beta$.
   Give examples for which $\alpha < \beta$ and $\alpha = \beta$.

4. If $x_n \leqslant a_n \leqslant y_n$ for all n and both (x_n) and (y_n) converge to l, prove that (a_n) also converges to l.

5. Prove that if $x_n \to \alpha$ as $n \to \infty$, then $|x_n| \to |\alpha|$ as $n \to \infty$. Show also that the converse is false unless $\alpha = 0$.

6. Find the limits as $n \to \infty$ of

   (a) $\dfrac{1 + 2^{1/2} + 3^{1/3} + \cdots + n^{1/n}}{n}$ (b) $\dfrac{(n+1)}{\sqrt[n]{n!}}$

7. By considering $s_n = (-1)^n$, show that the converse of Cauchy's theorem on limits is false.

5.5 Monotonic sequences

If (x_n) is a sequence for which $x_n \leqslant x_{n+1}$ for all n, then the sequence is said to be *increasing*. If equality is not allowed the sequence is said to be *strictly increasing*. When $x_n \geqslant x_{n+1}$ for all n, the sequence is *decreasing; strictly decreasing* if equality is not allowed. A sequence which is either increasing or decreasing is called *monotonic*.

To decide whether or not a sequence (x_n) is monotonic consider either $x_n - x_{n+1}$ or x_n/x_{n+1}. In the former case the difference must be of constant sign and in the latter case the ratio must be compared with unity.

Example 1 If $x_n = (2n-7)/(3n+2)$, then

$$x_n - x_{n+1} = \frac{-25}{(3n+2)(3n+5)} < 0 \qquad \text{for all } n.$$

Therefore (x_n) is increasing.

Example 2 If $x_n = n/(n^2+1)$, then

$$\begin{aligned}\frac{x_n}{x_{n+1}} &= 1 + \frac{n^2+n-1}{(n+1)(n^2+1)}\\ &> 1 \qquad \text{for } n = 1, 2, 3, \ldots .\end{aligned}$$

That is (x_n) is increasing.

Theorem If (x_n) is increasing, then either $\sup x_n = M$ and $x_n \to M$ as $n \to \infty$ or (x_n) is unbounded and $x_n \to +\infty$ as $n \to \infty$.

Similarly a decreasing sequence either converges or diverges to $-\infty$. In other words a *monotonic sequence cannot oscillate*.

Proof. Suppose $\sup x_n = M$. Then, given any $\epsilon > 0$,

$$\begin{aligned} x_n &\leqslant M < M+\epsilon && \text{for all } n,\\ &> M-\epsilon && \text{for at least one value of } n. \end{aligned}$$

Let this value be N; that is $x_N > M-\epsilon$. Since (x_n) is increasing, $x_n > x_N$ for $n > N$. Since

$$M-\epsilon < x_n < M+\epsilon \qquad \text{for } n > N,$$

(x_n) converges to M.

If (x_n) has no supremum, then x_n is unbounded. That is, for any $K > 0, \exists N.\, x_N > K$. Since (x_n) is increasing $x_n > x_N > K$ for $n > N$, which means $x_n \to \infty$ as $n \to \infty$.

Corollary If (x_n) is increasing and $x_n < K$ for all n, then $x_n \to l \leqslant K$ as $n \to \infty$.

Example 1 A sequence (x_n) is defined by the relation

$$x_1 = 1, \quad x_{n+1} = \sqrt{(x_n+1)}, n = 1, 2, 3, \ldots .$$

Then

$$x_n \to \tfrac{1}{2}(1+\sqrt{5}) \qquad \text{as } n \to \infty.$$

The first few terms of the sequence are $1, \sqrt{2}, \sqrt{(1+\sqrt{2})}$, $\sqrt{[1+\sqrt{(1+\sqrt{2})}]}$. We first show that the sequence is monotonic.

$$\begin{aligned} x_{n+1}-x_n &= \sqrt{(x_n+1)}-\sqrt{(x_{n-1}+1)} \qquad (n = 2, 3, 4, \ldots)\\ &= \frac{x_n - x_{n-1}}{\sqrt{(x_n+1)}+\sqrt{(x_{n-1}+1)}}. \end{aligned}$$

The radical sign indicates the positive square root, and so $x_{n+1} \gtreqless x_n$ according as $x_n \gtreqless x_{n-1}$, $n = 2, 3, \ldots$.

Now $x_2 = \sqrt{2} > x_1$, therefore $x_3 > x_2$, and by induction, $x_{n+1} > x_n$, $n = 1, 2, 3, \ldots$. That is (x_n) is an increasing sequence.

Again

$$\begin{aligned} x_{n+1}^2 - x_n^2 &= (x_n+1) - x_n^2 \qquad (n = 1, 2, 3, \ldots)\\ &= (\alpha - x_n)(1/\alpha + x_n), \end{aligned}$$

where α is the positive root of $x^2 = x + 1$; that is $\alpha = \frac{1}{2}(1 + \sqrt{5})$. Since (x_n) is increasing, the left-hand side is positive and $x_n < \alpha, n = 1, 2, 3, \ldots$ Thus (x_n) is an increasing sequence which is bounded above and so, by the Corollary,

$$x_n \to l \leqslant \alpha \qquad \text{as } n \to \infty.$$

Since $x_{n+1} = \sqrt{(x_n + 1)}$ and (x_n) converges to l, then l must be given by $l = \sqrt{(l + 1)}$ (using Property 2, Section 5.4). That is $l^2 = l + 1$, hence $l = \alpha$ or $-1/\alpha$.

But $x_n > 0$ for all n and so $\lim_{n \to \infty} x_n \geqslant 0$, and we have established that the sequence (x_n) converges to $\alpha = \frac{1}{2}(1 + \sqrt{5})$.

Example 2 If $x_{n+1} = x_n^2 + \frac{1}{4}, n = 1, 2, 3, \ldots$ and $x_1 = a$, then (x_n) is an increasing sequence. If $0 < a \leqslant \frac{1}{2}$, then $x_n \to \frac{1}{2}$ as $n \to \infty$.

We have

$$x_{n+1} - x_n = x_n^2 + \tfrac{1}{4} - x_n = (x_n - \tfrac{1}{2})^2 \geqslant 0 \qquad \text{for all } n.$$

Therefore (x_n) is increasing. Also

$$\begin{aligned} x_{n+1} - \tfrac{1}{2} &= (x_n^2 + \tfrac{1}{4}) - \tfrac{1}{2} \\ &= (x_n - \tfrac{1}{2})(x_n + \tfrac{1}{2}), \qquad n = 1, 2, 3, \ldots. \end{aligned}$$

Then $x_{n+1} \gtreqless \frac{1}{2}$ according as $x_n \gtreqless \frac{1}{2}, n = 1, 2, \ldots,$ and, by induction $x_n \gtreqless \frac{1}{2}$ according as $x_1 \gtreqless \frac{1}{2}$. Here $x_1 = a$; so if $0 < a < \frac{1}{2}, x_1 < \frac{1}{2}$ and $x_n < \frac{1}{2}$. That is (x_n) is an increasing sequence bounded above by $\frac{1}{2}$, and so

$$x_n \to l \leqslant \tfrac{1}{2} \text{ as } n \to \infty,$$

where l is given by $l = l^2 + \frac{1}{4}$. When $x_1 = a = \frac{1}{2}$, then $x_n = \frac{1}{2}$ for all n, therefore, when $0 < a \leqslant \frac{1}{2}$, (x_n) converges to $\frac{1}{2}$.

If $a > \frac{1}{2}$, then $x_n > \frac{1}{2}$; Let K be any positive number greater than $\frac{1}{2}$. Then $x_n^2 - K = x_n - \frac{1}{4} - K$, and so $x_n > K$ implies that $x_{n+1} > K + \frac{1}{4} > K$. That is (x_n) is strictly increasing and unbounded;

$$x_n \to \infty \qquad \text{as } n \to \infty.$$

Note. The assumption that x_n has a limit l used in the equation defining x_n gives an equation for the possible value of l *if it exists*. In Example 2 we have seen that $l = \frac{1}{2}$ *when it exists*, that is when $0 < a \leqslant \frac{1}{2}$, but when $a > \frac{1}{2}$ the limit does not exist. This example shows the importance of establishing the *existence* of the limit before attempting to find its value.

EXERCISES 5(c)

1. Determine which of the following sequences are monotonic:

(a) $\dfrac{3n+2}{2n-5}$ (b) $\dfrac{n^2+1}{n}$ (c) $\dfrac{2n^2-1}{2n^2+1}$

(d) $1+\dfrac{1}{n^2}$ (e) $\dfrac{(n-1)(n+2)}{(n+1)(n-3)}$ (f) $1+\dfrac{(-1)^n}{n}$

(g) $n+(-1)^n$ (h) $2n+(-1)^n$ (i) $\dfrac{10^n}{n!}$.

2. Prove that if $x_{n+1} \geqslant Kx_n$ for $n \geqslant N$ and $K > 1$, then

$$x_n \to \infty \quad n \to \infty.$$

3. Prove that if $x_{n+1} \leqslant Kx_n$ for $n \geqslant N$ and $0 < K < 1$, then

$$x_n \to 0 \quad \text{as } n \to \infty.$$

4. Prove that if $x_n > 0$ and $\lim_{n\to\infty} x_n/x_{n+1} = l$, then

(a) $x_n \to \infty$ as $n \to \infty$ if $0 < l < 1$,

(b) $x_n \to 0$ as $n \to \infty$ if $l > 1$.

5. Show that

(a) $(x^n/n!)$ is a null sequence for all real x.

(b) $\{\binom{m}{n}x^n\}$ is null for $|x| < 1$,

$$\text{where} \binom{m}{n} = \frac{m(m-1)(m-2)\cdots(m-n+1)}{n!}.$$

6. Given that $a_n = (1+1/n)^n$ and $b_n = (1-1/n)^{-n}$, prove that

(a) (a_n) is increasing

(b) (b_n) is decreasing

(c) $b_n > a_n$

(d) $(b_n - a_n)$ is null

(e) if $\lim_{n\to\infty} a_n = e$, then $2 < e \leqslant 3$.

7. Prove that if (x_n) is a bounded sequence of positive numbers, then

$$\frac{x_n}{x_1+x_2+\cdots+x_n} \to 0 \qquad \text{as } n \to \infty.$$

8. Prove that if (a_n) is a monotonic sequence and $b_n = (a_1 + a_2 + a_3 + \cdots + a_n)/n$, then b_n is also monotonic.
By taking $a_n = a^{n-1}(1-a)$, show that the sequence $\{(1-a^n)/n\}$ is decreasing when $0 < a < 1$. Deduce that

$$na^{n-1}(1-a) < 1 - a^n < n(1-a), \qquad 0 < a < 1.$$

9. If $u_{n+1} = \sqrt{(3u_n)}$, $u_1 = 1$ show that (u_n) is increasing and bounded, and deduce that $u_n \to 3$ as $n \to \infty$.

10. A sequence (u_n) is defined by the relation

$$u_{n+1} = \frac{6(1+u_n)}{7+u_n}, \qquad u_1 = c > 0.$$

Show that (u_n) is increasing and bounded above if $c < 2$; find $\lim_{n\to\infty} u_n$ in this case. What happens if $c \geqslant 2$?

11. Given $x_{n+1} = \sqrt{(a + x_n)}$, $a > 0$ and $x_1 > 0$, prove that $x_n \to \alpha$ as $n \to \infty$, where α is the positive root of $x^2 = x + a$.

12. Show that, if $a \geqslant 1$, $0 < x_0 < a^2$ and the positive square root is taken, $x_{n+1} = a - \sqrt{(a^2 - x_n)}$, $n = 0, 1, 2, \ldots$ defines a sequence of positive numbers (x_n) such that

(a) x_n is decreasing and null, and

(b) $x_n/x_{n+1} \to 2a$ as $n \to \infty$.

5.6 Subsequences

The sequence (y_n) is called a *subsequence* of the sequence (x_n) if all the terms of (y_n) are taken *in order* from those of (x_n). For example

$$1, \tfrac{1}{2}, (\tfrac{1}{2})^2, (\tfrac{1}{2})^3, \ldots, (\tfrac{1}{2})^n, \ldots \text{ is a subsequence of } 1, \tfrac{1}{2}, \tfrac{1}{3}, \tfrac{1}{4}, \ldots, \tfrac{1}{n}, \ldots.$$

Theorem Every bounded sequence contains at least one convergent subsequence.

We use the method of repeated bisection to prove this theorem. Suppose $a \leqslant x_n \leqslant b$ for all n. Bisect the interval $[a, b]$; since (x_n) is an infinite sequence there is an infinite number of terms in at least one of the subintervals. Choose the subinterval containing the infinite number of terms, or if both do, choose the right-hand one. Call this interval $[a_1, b_1]$. In both cases

$$a \leqslant a_1 < b_1 \leqslant b \quad \text{and} \quad b_1 - a_1 = \tfrac{1}{2}(b-a).$$

Now bisect the interval $[a_1, b_1]$ and choose the subinterval $[a_2, b_2]$ containing infinitely many terms of the sequence; if both halves do choose the right hand one. Then

$$a \leqslant a_1 \leqslant a_2 < b_2 \leqslant b_1 \leqslant b, \quad \text{and} \quad b_2 - a_2 = \tfrac{1}{2}(b_1 - a_1) = (\tfrac{1}{2})^2(b-a).$$

Continuing this process of bisection we obtain a sequence of intervals $[a_n, b_n]$ in which (a_n) is increasing, (b_n) decreasing and

(a) $a \leqslant a_n < b_n \leqslant b$ for all n;

(b) $b_n - a_n = (\tfrac{1}{2})^n (b-a)$;

(c) $[a_n, b_n]$ contains infinitely many terms of the sequence (x_n).

Since (a_n) and (b_n) are monotonic and bounded,

$$a_n \to l \quad \text{and} \quad b_n \to l' \quad \text{as } n \to \infty,$$

and from (b) $l = l'$. Hence, for any $\epsilon > 0$,

$$|x_n - l| < \epsilon \qquad \text{for an infinity of values of } n.$$

In particular, if $\epsilon = 1$, then for a suitable $n = \nu_1$ (say)

$$|x_{\nu_1} - l| < 1.$$

Now take $\epsilon = \tfrac{1}{2}$, then $\exists n = \nu_2 > \nu_1$. $|x_{\nu_2} - l| < \tfrac{1}{2}$. In this way we can take a sequence of values of $\epsilon = 1/k$ and find an increasing sequence of integers ν_k such that

$$|x_{\nu_k} - l| < \frac{1}{k}, \qquad k = 1, 2, 3, \ldots .$$

Thus we have found a subsequence of (x_n) which converges to l.

From this theorem we can deduce the following.

Corollary If (x_n) is a bounded sequence, then either (x_n) converges or it contains two subsequences which converge to different limits.

Proof. Since (x_n) is bounded there exists a subsequence (x_{ν_k}) which converges to l. Consider $|l - x_n| \geqslant \epsilon$ for any $\epsilon > 0$; either this holds for only a finite number of terms of (x_n) (or none) or for some $\epsilon > 0$ there is an infinity of terms of (x_n) for which the inequality holds. The first case implies

$$|l - x_n| < \epsilon \qquad \text{for } n > n_0,$$

where n_0 is a number larger than the index of each term satisfying $|l - x_n| \geqslant \epsilon$. That is, (x_n) converges to l.

In the second case, suppose that for a particular $\epsilon = \epsilon_1$, $|l - x_n| \geqslant \epsilon_1$ for an infinity of values of n. So there is a subsequence (x_{μ_k}) such that

$$|l - x_{\mu_k}| > \epsilon_1 \qquad \text{for all } k.$$

Since (x_{μ_k}) is bounded, it contains a subsequence $(x_{\mu'_k})$ which converges to l', and

$$|l' - x_{\mu'_k}| < \tfrac{1}{2}\epsilon_1 \qquad \text{for all } k.$$

Also since $(x_{\mu'_k})$ is a subsequence of (x_{μ_k}), then

$$|l - x_{\mu'_k}| > \epsilon \qquad \text{for all } k.$$

Thus $|l - l'| = |(l - x_{\mu'_k}) - (l' - x_{\mu'_k})|$

$$\geqslant |l - x_{\mu'_k}| - |l' - x_{\mu'_k}|$$

$$\geqslant \epsilon_1 - \tfrac{1}{2}\epsilon_1 = \tfrac{1}{2}\epsilon_1 > 0.$$

Therefore $l' \neq l$.

5.7 Cauchy's general principle of convergence

The definition of convergence given in Section 5.3 requires us to know the limit to which the sequence converges. We now consider Cauchy's general principle of convergence which does not contain the limiting value of x_n.

Theorem A necessary and sufficient condition for the convergence of (x_n) is that

$$\forall \epsilon > 0; \exists m.\ |x_{m+p} - x_m| < \epsilon \qquad \text{for } p = 1, 2, 3, \ldots.$$

A sequence which satisfies this condition is called a *Cauchy sequence*.

Proof. The condition is necessary, for if (x_n) converges to l then

$$\forall \epsilon > 0; \exists m.\ |l - x_n| < \tfrac{1}{2}\epsilon \qquad \text{for } n \geqslant m.$$

Consequently,

$$|x_{m+p} - x_m| = |(l - x_m) - (l - x_{m+p})|$$

$$\leqslant |l - x_m| + |l - x_{m+p}|$$

$$< \epsilon \qquad \text{for } p = 1, 2, 3, \ldots.$$

Now suppose $|x_{m+p} - x_m| < \frac{1}{4}\epsilon$ for $p = 1, 2, 3, \ldots$; then certainly the sequence is bounded, and if *not* convergent it must contain at least two subsequences converging to different limits l and l'. Then, there are integers p and q (which, of course, depend on ϵ) such that

$$|l - x_{m+p}| < \tfrac{1}{4}\epsilon \quad \text{and} \quad |l' - x_{m+q}| < \tfrac{1}{4}\epsilon.$$

Then

$$\begin{aligned} |l - l'| &= |(l - x_{m+p}) + (x_{m+p} - x_m) - (x_{m+q} - x_m) - (l' - x_{m+q})| \\ &\leqslant |l - x_{m+p}| + |x_{m+p} - x_m| + |x_{m+q} - x_m| + |l' - x_{m+q}| \\ &< \epsilon. \end{aligned}$$

But, since ϵ is arbitrary, this implies $l' = l$, which contradicts our assumption that $l' \neq l$.

Therefore (x_n) converges, and the sufficiency of the condition is proved.

Notes.

1. A necessary and sufficient condition for the convergence of (x_n) is that

$$\forall \epsilon > 0; \exists N.\ |x_n - x_m| < \epsilon \qquad \text{for all } n > m \geqslant N.$$

2. It is *not* sufficient for $\lim_{n \to \infty} (x_{n+p} - x_n) = 0$ for a *fixed* p. For example $(\log n)$ is divergent, but

$$\lim_{n \to \infty} [\log (n + p) - \log n] = \lim_{n \to \infty} \log (1 + p/n) = 0$$

if p is fixed.

5.8 Upper and lower limits

For any bounded sequence (x_n), suppose there is a number Λ with the properties

$$\begin{aligned} \forall \epsilon > 0; \exists n_0.\ x_n &< \Lambda + \epsilon \qquad \text{for } n > n_0, \quad \text{and} \\ x_n &> \Lambda - \epsilon \qquad \text{for an infinity of values of } n. \end{aligned}$$

Then Λ is called the *upper limit* or *limit superior* of the sequence and is written

$$\Lambda = \overline{\lim_{n \to \infty}}\, x_n = \lim_{n \to \infty} \sup x_n.$$

If (x_n) is not bounded above as we take $\Lambda = +\infty$.

Similarly, λ, the *lower limit* or *limit inferior* of the bounded sequence (x_n) has the properties

$$\begin{aligned} \forall \epsilon > 0; \exists n_0.\ x_n &> \lambda - \epsilon \qquad \text{for } n > n_0 \quad \text{and} \\ x_n &< \lambda + \epsilon \qquad \text{for an infinity of values of } n. \end{aligned}$$

We write

$$\lambda = \underline{\lim}_{n \to \infty} x_n = \lim_{n \to \infty} \inf x_n .$$

If (x_n) is not bounded below we take $\lambda = -\infty$.

For a sequence whose lower and upper bounds are m and M we have

$$m \leqslant \lambda \leqslant \Lambda \leqslant M.$$

Also $\lim_{n \to \infty} x_n = l$ if and only if $\lambda = \Lambda = l$. The proof of this is left as an exercise.

EXERCISES 5(d)

1. Find the lower and upper limits of the following sequences:

(a) $0, 1, -1, 0, 1, -1, \ldots$ (b) $-\frac{3}{2}, \frac{1}{2}, -\frac{4}{3}, \frac{1}{3}, -\frac{5}{4}, \frac{1}{4}, \ldots$

(c) $(-1)^n n$ (d) $\{1 + (-1)^n\}/3$ (e) $(-1)^n n/(2n+1)$

(f) $(-1)^n + 1/n$ (g) $(-1)^n \operatorname{th} n$ (h) $(1 + 1/n) \sin \frac{1}{4} n\pi$.

2. Prove that if (x_n) is a bounded sequence and $M(n) = \sup_{r \geqslant n} x_r$, then $M(n)$ is monotonic decreasing and that

$$\lim_{n \to \infty} M(n) = \Lambda = \overline{\lim}_{n \to \infty} x_n .$$

3. Prove that if $m(n) = \inf_{r \geqslant n} x_r$, where (x_n) is a bounded sequence, then $\lim_{n \to \infty} m(n) = \lambda = \underline{\lim}_{n \to \infty} x_n$.

4. Prove that if (x_n) and (y_n) are two sequences of real numbers and $x_n > y_n$ for all n, then $\overline{\lim}_{n \to \infty} x_n \geqslant \overline{\lim}_{n \to \infty} y_n$. Is the result true if $x_n > y_n$ for an infinity of values of n? Show that in this case

$$\overline{\lim}_{n \to \infty} x_n \geqslant \underline{\lim}_{n \to \infty} y_n .$$

5. Prove that if (x_n) is a bounded sequence of real numbers, then

$$\overline{\lim}_{n \to \infty} = n(x_{n+1} - x_n) \geqslant 0.$$

6. Prove that if (x_n) is a sequence of positive numbers, then

$$\underline{\lim}_{n \to \infty} \frac{x_{n+1}}{x_n} \leqslant \underline{\lim}_{n \to \infty} x_n^{1/n} \leqslant \overline{\lim}_{n \to \infty} x_n^{1/n} \leqslant \overline{\lim}_{n \to \infty} \frac{x_{n+1}}{x_n} .$$

7. Given $x_n = (2 + \cos n\pi)^n$, find the upper and lower limits of (x_{n+1}/x_n) and $(x_n^{1/n})$.

8. Prove that a Cauchy sequence is bounded.

9. Prove that $\lim_{n \to \infty} x_n = l$ if and only if $\lambda = \Lambda = l$.

5.9 Continuity in terms of sequences

Theorem A necessary and sufficient condition for a function $f(x)$ to be continuous at $x = a$ is that, for every sequence (x_n) converging to a, the sequence $\{f(x_n)\}$ converges to $f(a)$.

Proof. If $f(x)$ is continuous at $x = a$, then

$$\forall \epsilon > 0; \exists \delta.\ |f(x) - f(a)| < \epsilon \qquad \text{for } |x - a| < \delta.$$

Since (x_n) converges to a,

$$\exists N.\ |x_n - a| < \delta \qquad \text{for } n > N.$$

Then

$$|f(x_n) - f(a)| < \epsilon \qquad \text{for } n > N;$$

that is

$$f(x_n) \to f(a) \qquad \text{as } n \to \infty.$$

To prove the sufficiency of the condition, we assume $f(x)$ to be discontinuous at $x = a$. Then $\exists\, \epsilon > 0$, $\delta > 0$ such that for some x,

$$|x - a| < \delta \text{ and } |f(x) - f(a)| \geqslant \epsilon.$$

Now take $\delta = 1/n$ and construct a sequence (x_n) (see Section 5.6) which converges to a and satisfies

$$|x_n - a| < \frac{1}{n} \quad \text{and} \quad |f(x_n) - f(a)| \geqslant \epsilon.$$

But this contradicts the condition $\{f(x_n)\}$ converges to $f(a)$; therefore $f(x)$ is continuous at $x = a$.

In Section 3.5 we proved, using the idea of supremum, the following theorem which we are now in a position to prove in terms of sequences.

Theorem If a function f is continuous in $[a, b]$, then f is bounded there.

Proof. Suppose the theorem is false. Then, for any positive integer n there is a term of a sequence (x_n) of points of $[a, b]$, that is $a \leqslant x_n \leqslant b$, such that $|f(x_n)| > n$.

Since (x_n) is bounded it contains a convergent subsequence, say (x_{ν_n}), converging to a point ξ in $[a, b]$.

Now f is continuous in $[a, b]$ and so $\{f(x_{\nu_n})\}$ converges to $f(\xi)$. But this contradicts the assumed unboundedness of f and so proves the theorem.

5.10 Uniform continuity

Theorem A function $f(x)$ continuous on a closed interval is uniformly continuous there. (See Section 3.7.)

Proof. Suppose $f(x)$ is continuous on $[a, b]$ but not uniformly continuous there. Then there is an $\epsilon > 0$ and a pair of points x, x' in the interval, such that for $\delta > 0$

$$|x - x'| < \delta \quad \text{and} \quad |f(x) - f(x')| \geqslant \epsilon.$$

Now take $\delta = 1/n$; then for any positive integer n there exist points x_n, x_n' of $[a, b]$ such that

$$|x_n - x_n'| < \frac{1}{n} \quad \text{and} \quad |f(x_n) - f(x_n')| \geqslant \epsilon.$$

The sequence (x_n) is bounded and so contains a subsequence (x_{ν_k}) converging to a point ξ of $[a, b]$. Let the corresponding subsequence of (x_n') be (x_{ν_k}'). Then

$$\begin{aligned} |x_{\nu_k}' - \xi| &\leqslant |x_{\nu_k}' - x_{\nu_k}| + |x_{\nu_k} - \xi| \\ &< \frac{1}{\nu_k} + |x_{\nu_k} - \xi| \\ &\to 0 \qquad \text{as } k \to \infty. \end{aligned}$$

Therefore (x_{ν_k}') converges to ξ also. Now $f(x)$ is continuous at ξ, so $f(x_{\nu_k}) \to f(\xi)$ and $f(x_{\nu_k}') \to f(\xi)$ as $k \to \infty$.

But the definitions of x_{ν_k} and x_{ν_k}' ensure that

$$|f(x_{\nu_k}) - f(x_{\nu_k}')| \geqslant \epsilon > 0.$$

Therefore

$$\lim_{k \to \infty} |f(x_{\nu_k}) - f(x_{\nu_k}')| \geqslant \epsilon;$$

that is

$$|f(\xi) - f(\xi)| \geqslant \epsilon > 0,$$

which is false. Therefore $f(x)$ is uniformly continuous on $[a, b]$.

CHAPTER 6

Infinite Series

6.1 The sum of an infinite series

When the terms of a sequence (u_n) are added together in order we obtain an *infinite series*, namely

$$u_1 + u_2 + u_3 + \cdots + u_n + \cdots .$$

The series may be written

$$\sum_{n=1}^{\infty} u_n \quad \text{or} \quad \sum u_n .$$

Closely related to the series is the sequence of *partial sums* (s_n), which is formed by adding a finite number of terms of the series in order as follows:

$$s_1 = u_1, s_2 = u_1 + u_2, s_3 = u_1 + u_2 + u_3$$

and generally

$$s_n = u_1 + u_2 + u_3 + \cdots + u_n = \sum_{r=1}^{n} u_r .$$

The expression Σu_n always denotes the *infinite* series; consequently it is important to show the limits of summation when considering only a finite number of terms. The series Σu_n is said to be *convergent* or *divergent* according as the sequence of partial sums (s_n) is convergent or divergent. If $s_n \to s$ as $n \to \infty$, we say that the series Σu_n *converges to the sum* s.

Example 1 $1 + \frac{1}{2} + \frac{1}{4} + \frac{1}{8} + \cdots$.

Here $s_n = 1 + \frac{1}{2} + (\frac{1}{2})^2 + (\frac{1}{2})^3 + \cdots + (\frac{1}{2})^n$

$= 2\{1 - (\frac{1}{2})^n\}$ (using the formula for the sum of a geometric progression)

$\to 2$ as $n \to \infty$,

and so the geometric progression converges to the sum 2.

Example 2 $1+3+5+\cdots$.

In this case the series diverges, for

$$\begin{aligned} s_n &= 1+3+5+\cdots+(2n-1) \\ &= \tfrac{1}{2}n\{2+2(n-1)\} \text{ (using the formula for the sum of an arithmetic progression)} \\ &= n^2 \end{aligned}$$

and $s_n \to \infty$ as $n \to \infty$.

Example 3 $1-3+5-\cdots$.

$$\begin{aligned} \text{Here } s_{2n} &= 1-3+5-7+\cdots+(4n-3)-(4n-1) \\ &= [1+5+9+\cdots+(4n-3)]-[3+7+11+\cdots+(4n-1)] \\ &= n(2n-1)-n(2n+1) \quad \text{(since each series in square brackets is an arithmetic progression)} \\ &= -2n. \end{aligned}$$

Also $s_{2n+1} = s_{2n}+(4n+1) = 2n+1$. Therefore the series oscillates infinitely.

Example 4 $\dfrac{1}{2.5}+\dfrac{1}{5.6}+\dfrac{1}{8.11}+\cdots$.

$$\begin{aligned} s_n &= \sum_{r=1}^{n} \frac{1}{(3r-1)(3r+2)} = \frac{1}{3}\sum_{r=1}^{n}\left(\frac{1}{3r-1}-\frac{1}{3r+2}\right) \\ &= \frac{1}{3}\left(\frac{1}{2}-\frac{1}{5}+\frac{1}{5}-\frac{1}{8}+\frac{1}{8}-\frac{1}{11}+\cdots\right. \\ &\qquad \left.+\frac{1}{3n-4}-\frac{1}{3n-1}+\frac{1}{3n-1}-\frac{1}{3n+2}\right) \\ &= \frac{1}{3}\left(\frac{1}{2}-\frac{1}{3n+2}\right) \\ &\to \tfrac{1}{6} \qquad \text{as } n \to \infty. \end{aligned}$$

Therefore the series $\Sigma\, 1/[(3n-1)(3n+2)]$ converges to $\frac{1}{6}$.

Example 5 $\sum\limits_{n=1}^{\infty}(x^{n-1}-x^n)$.

$$s_n = \sum_{r=1}^{n}(x^{r-1}-x^r) = x-x^n.$$

Therefore the series converges to x if $|x| < 1$, and diverges if $|x| > 1$. When $x = 1$ the series converges to zero, for then each term vanishes; when $x = -1$ the series diverges (oscillates finitely).

Example 6 $\sum_{n=1}^{\infty} \frac{n}{(n+1)!}$ converges to 1.

For

$$s_n = \sum_{r=1}^{n} \frac{r}{(r+1)!} = \sum_{r=1}^{n} \left\{\frac{1}{r!} - \frac{1}{(r+1)!}\right\}$$

$$= 1 - \frac{1}{(n+1)!}$$

$$\to 1 \qquad \text{as } n \to \infty.$$

6.2 Some properties of infinite series

Property 1 The convergence or divergence of an infinite series corresponds to the limiting behaviour of the sequence of partial sums, and consequently is not affected by changing a *finite* number of terms. Of course, if a finite number of terms are added to or subtracted from the series the sum, if the series converges, is changed.

Example $1 + \frac{1}{16} + \frac{1}{32} + \frac{1}{64} + \cdots$ is the series in Example 1 above with three terms omitted. Therefore it still converges, but its sum is no longer 2; in fact it is $2 - (\frac{1}{2} + \frac{1}{4} + \frac{1}{8}) = 1\frac{1}{8}$.

Property 2 If Σu_n and Σv_n are both convergent, then so also is $\Sigma(u_n \pm v_n)$.

Let the partial sums of Σu_n be (s_n) and those of Σv_n be (t_n). Then this property is equivalent to the theorem (Section 5.4) that if, as $n \to \infty$, $s_n \to s$ and $t_n \to t$, then $s_n \pm t_n \to s \pm t$ as $n \to \infty$.

Property 3 If Σu_n is convergent and Σv_n is divergent, then $\Sigma(u_n \pm v_n)$ is divergent.

This follows directly from a consideration of the behaviour of the partial sums of the series.

Property 4 If both Σu_n and Σv_n diverge, nothing can be inferred about the behaviour of $\Sigma(u_n \pm v_n)$. For example:

(a) $s_n = n$, $t_n = n^2$; then $s_n + t_n \to \infty$ and $s_n - t_n \to -\infty$.

(b) $s_n = n$, $t_n = (-1)^n - n$; then $s_n + t_n = (-1)^n$, which oscillates finitely.

(c) $s_n = n + 1/n$, $t_n = n$ then $s_n - t_n = 1/n$, which converges to zero.

Property 5 The terms of a series may be multiplied by a non-zero constant without affecting the convergence or divergence of the series. That is, if $k \neq 0$, then Σu_n and $\Sigma(ku_n)$ either both converge or both diverge, since $(ks_n) = k(s_n)$.

Property 6 If Σu_n is convergent, then so is the series formed by grouping terms in brackets *without altering their order*, and the two series have the same sum.

For the partial sums of the series with brackets is a subsequence of (s_n); and if (s_n) converges to s then so does any subsequence of (s_n).

It is important that the order of the terms of Σu_n should not be altered, because a change in order can change the sum or even make a convergent series diverge (Section 7.5). For example, if

$$s = 1 - \tfrac{1}{2} + \tfrac{1}{3} - \tfrac{1}{4} + \cdots ,$$

then

$$\begin{aligned} s + \tfrac{1}{2}s &= 1 - \tfrac{1}{2} + \tfrac{1}{3} - \tfrac{1}{4} + \cdots + \tfrac{1}{2}(1 - \tfrac{1}{2} + \tfrac{1}{3} - \tfrac{1}{4} + \cdots) \\ &= 1 + \tfrac{1}{3} - \tfrac{1}{2} + \tfrac{1}{5} + \tfrac{1}{7} - \tfrac{1}{4} + \cdots . \end{aligned}$$

This series contains exactly the same terms as the original series but in a different order.

Note. The removal of brackets from a series may also change its behaviour. For example

$$(1 - 1) + (1 - 1) + (1 - 1) + \cdots$$

converges to zero, but

$$1 - 1 + 1 - 1 + 1 - 1 + 1 - 1 + \cdots$$

diverges (oscillates finitely).

6.3 The general principle of convergence

Theorem A necessary and sufficient condition that the series Σu_n should converge is

$$\forall \epsilon > 0; \exists m. \left| \sum_{r=m+1}^{m+p} u_n \right| < \epsilon \qquad \text{for } p = 1, 2, 3, \ldots ,$$

or in terms of partial sums, s_n,

$$\forall \varepsilon > 0; \exists m.\ |s_{m+p} - s_m| < \epsilon \qquad \text{for } p = 1, 2, 3, \ldots .$$

Proof. This was proved in Section 5.7.

We can also write the general principle of convergence in the form,

$$\forall \epsilon > 0; \exists N.\ |s_n - s_m| < \epsilon \qquad \text{for } n > m \geqslant N.$$

Notice that

$$|s_n - s_m| = \left| \sum_{r=m+1}^{n} u_r \right| .$$

Corollary If Σu_n is convergent, then $u_n \to 0$ as $n \to 0$.

Proof. For $u_n = s_n - s_{n-1}$, and the result follows from the above theorem by taking $m = n - 1$ and $p = 1$.

Notice that this corollary gives a necessary but *not sufficient* condition for convergence. For example, if $u_n = \sqrt{n} - \sqrt{(n-1)}$, then

$$\begin{aligned} s_n &= 1 + (\sqrt{2} - 1) + (\sqrt{3} - \sqrt{2}) + \cdots + (\sqrt{n} - \sqrt{(n-1)}) \\ &= \sqrt{n}, \end{aligned}$$

and so Σu_n diverges. But

$$u_n = \sqrt{n} - \sqrt{(n-1)} = \frac{1}{\sqrt{n} + \sqrt{(n-1)}} \to 0 \qquad \text{as } n \to \infty.$$

This property is a very useful *test of divergence*:

If $u_n \not\to 0$ as $n \to \infty$, then Σu_n diverges.

EXERCISES 6(a)

1. Find the sum of the first n terms of the following series and hence decide whether each series converges or diverges. If the series converges find its sum:

(a) $\frac{1}{3} + \frac{2}{9} + \frac{4}{27} + \cdots$ (b) $2 - 4 + 6 - 8 + \cdots$

(c) $\dfrac{1}{1.2} + \dfrac{1}{2.3} + \dfrac{1}{3.4} + \cdots$ (d) $\dfrac{1}{1.2.3} + \dfrac{1}{2.3.4} + \dfrac{1}{3.4.5} + \cdots$

(e) $\log 2 + \log \frac{3}{2} + \log \frac{4}{3} + \cdots$ (f) $x + \frac{1}{2} + x^2 + \frac{1}{4} + x^3 + \frac{1}{8} + \cdots$

(g) $\displaystyle\sum_{r=1}^{\infty} \frac{2r+1}{r^2(r+1)^2}$ (h) $\displaystyle\sum_{r=0}^{\infty} \frac{(-1)^r (2r+3)}{(r+1)(r+2)}$

(i) $\sum_{r=0}^{\infty} \dfrac{x}{(1+x)^r}$ (j) $\sum_{r=1}^{\infty} \dfrac{x^2}{(1+x^2)^{r-1}}$.

2. Prove that, if d_n is increasing and unbounded, then

 (a) $\Sigma(d_{n+1} - d_n)$ is divergent and

 (b) $\Sigma(1/d_n - 1/d_{n+1})$ is convergent.

3. Prove that if Σa_n converges and $a_n \neq 0$, then $\Sigma(1/a_n)$ diverges.

6.4 Series of positive terms

The definition in Section 6.1 and the general principle of convergence in Section 6.3 are not usually very helpful when trying to decide whether or not a given series converges. This is because in both cases a knowledge of the partial sums is required, and generally it is not possible to find a simple formula for s_n. In the next few sections we shall develop some tests for convergence which do not require a knowledge of the partial sums.

We first restrict ourselves to series of non-negative terms, that is, series for which $u_n \geqslant 0$ for all n. For such series the sequence of partial sums (s_n) is increasing, since $s_{n+1} = s_n + u_{n+1} \geqslant s_n$. Hence Σu_n converges if s_n is bounded and diverges (to $+\infty$) if s_n is unbounded. (See Section 5.5.)

Example $\sum_{n=1}^{\infty} n^{-\alpha}$ converges if $\alpha > 1$ and diverges if $\alpha \leqslant 1$.

$\alpha > 1$. Consider

$$s_N = 1 + (2^{-\alpha} + 3^{-\alpha}) + (4^{-\alpha} + 5^{-\alpha} + 6^{-\alpha} + 7^{-\alpha}) + (8^{-\alpha} + \cdots + 15^{-\alpha}) + \cdots + (2^{-(n-1)\alpha} + \cdots + (2^n - 1)^{-\alpha}),$$

where $N = 2^n - 1$. Then

$$\begin{aligned} s_N &< 1 + 2.2^{-\alpha} + 4.4^{-\alpha} + 8.8^{-\alpha} + \cdots + 2^{n-1}.2^{-(n-1)\alpha} \\ &= 1 + a + a^2 + a^3 + \cdots + a^{n-1}, \text{ where } a = 2^{-(\alpha-1)} \\ &= \frac{1-a^n}{1-a} . \end{aligned}$$

Hence, since $a > 0$,

$$s_N < \frac{1}{1-2^{-(\alpha-1)}} ,$$

that is the partial sums s_N are bounded and so $\Sigma n^{-\alpha}$ converges when $\alpha > 1$.

$\alpha = 1$. Let $N = 2^n$; then

$$s_N = 1 + \tfrac{1}{2} + (\tfrac{1}{3} + \tfrac{1}{4}) + (\tfrac{1}{5} + \cdots + \tfrac{1}{8}) + \cdots + \left(\frac{1}{2^{n-1}+1} + \cdots + \frac{1}{2^n}\right)$$

$$> \tfrac{1}{2} + \tfrac{1}{2} + 2.\tfrac{1}{4} + 4.\tfrac{1}{8} + \cdots + 2^{n-1}.\tfrac{1}{2}n$$

That is $s_N > \frac{1}{2}(n+1)$, and so is unbounded. Therefore $\Sigma 1/n$ diverges. This important series is called the *harmonic series*.

$\alpha < 1$. Clearly if $\alpha \leqslant 0$, $n^{-\alpha} \nrightarrow 0$ as $n \to \infty$. Hence the series diverges when $\alpha \leqslant 0$. When $0 < \alpha < 1$,

$$s_n = 1 + 2^{-\alpha} + 3^{-\alpha} + \cdots + n^{-\alpha}$$

$$> 1 + \tfrac{1}{2} + \tfrac{1}{3} + \cdots + 1/n,$$

and we have proved that the partial sums of the harmonic series are unbounded. Therefore s_n is unbounded and $\Sigma n^{-\alpha}$ diverges when $\alpha < 1$.

6.5 Comparison test

Theorem (a) Suppose $u_n \geqslant 0$ and Σc_n is a convergent series of positive terms. If k is a fixed positive number and $u_n \leqslant kc_n$ for $n > n_0$, then Σu_n converges.

Proof. Since Σc_n is convergent, we have by the general principle of convergence

$$\forall \epsilon > 0; \exists m. \sum_{r=m+1}^{n} c_r < \frac{\epsilon}{k} \qquad \text{for } n > m.$$

Now

$$\sum_{r=m+1}^{n} u_r \leqslant k \sum_{r=m+1}^{n} c_r < \epsilon \qquad \text{for } n > m > n_0.$$

Therefore, by the general principle of convergence Σu_n converges.

Theorem (b) Suppose $u_n \geqslant 0$ and Σd_n is a divergent series of positive terms. If k is a fixed positive number and $u_n \geqslant kd_n$ for $n > n_0$, then Σu_n diverges.

Proof. Since Σd_n is divergent and $d_n > 0$, then

$$\forall M > 0; \exists N. \sum_{r=n+1}^{N} d_r > \frac{M}{k}.$$

Hence

$$\sum_{r=n_0+1}^{N} u_r \geqslant k \sum_{r=n_0+1}^{N} d_r > M,$$

and so Σu_n diverges.

Theorem (c) If Σu_n, Σv_n are two series of positive terms such that

$$\lim_{n\to\infty} \frac{u_n}{v_n} = l \neq 0,$$

then Σu_n and Σv_n converge or diverge together.

Proof. Using Property 1, Section 5.4, since $l > 0$,

$$\tfrac{1}{2}l < \frac{u_n}{v_n} < \tfrac{3}{2}l \qquad \text{for } n > n_0.$$

If Σv_n is convergent then Σu_n converges using (a), for $u_n < \frac{3}{2}lv_n$ for $n > n_0$, and Σv_n is convergent. If Σv_n is divergent then since $u_n > \frac{1}{2}lv_n$, Σu_n is divergent (using (b) above).

Similarly if Σu_n converges (diverges) so also does Σv_n. For another form of comparison test see Exercises 6(b), No. 3.

Example 1 $\Sigma n/(n^4 - 3)$ converges, for

$$u_n = \frac{n}{n^4 - 3} < \frac{n}{n^4 - \frac{1}{2}n^4} = \frac{2}{n^3} \qquad \text{if } n > 2,$$

and Σn^{-3} converges.

Usually it is easier to use the limit form of the comparison test. Taking $v_n = n^{-3}$, we have

$$\frac{u_n}{v_n} = \frac{n^4}{n^4 - 3} = \frac{1}{1 - 3n^{-4}} \to 1 \qquad \text{as } n \to \infty.$$

Example 2 $\Sigma n^2/(n^3 - 3)$

Here $u_n = n^2/(n^3 - 3)$ behaves like $n^2/n^3 = 1/n$ when n is large, so we take $v_n = 1/n$. Then

$$\frac{u_n}{v_n} = \frac{n^3}{n^3 - 3} \to 1 \qquad \text{as } n \to \infty,$$

and Σu_n diverges by comparison with Σn^{-1}.

Example 3 $\Sigma(3^n+1)/(4^n-1)$

Take $v_n = (\frac{3}{4})^n$; then

$$\frac{u_n}{v_n} = \frac{3^n+1}{4^n-1}\cdot\frac{4^n}{3^n} = \frac{1+(\frac{1}{3})^n}{1-(\frac{1}{4})^n} \to 1 \qquad \text{as } n\to\infty.$$

Therefore Σu_n converges by comparison with the convergent geometric progression $\Sigma(\frac{3}{4})^n$.

6.6 Cauchy's root test

Theorem (a) If $u_n > 0$ and $u_n^{1/n} \geqslant 1$, then Σu_n diverges.

(b) If $u_n > 0$ and there is a constant k such that $u_n^{1/n} \leqslant k < 1$, then Σu_n converges.

(c) *Limit form.* If $u_n > 0$ and $\lim_{n\to\infty} u_n^{1/n} = l$, then Σu_n converges if $l < 1$ and diverges if $l > 1$.

Proof. In the limit form of the test if $l > 1$ then $u_n^{1/n} > 1$ for $n > n_0$ and case (a) applies; if $l < 1$, choose a number k such that $l < k < 1$; then $u_n^{1/n} < k$ for $n > n_0$ and case (b) applies. We now prove cases (a) and (b).

If $u_n^{1/n} \geqslant 1$, then $u_n \geqslant 1$ and $u_n \not\to 0$ as $n\to\infty$, therefore Σu_n diverges.

If $u_n^{1/n} \leqslant k < 1$, then $u_n \leqslant k^n$ and Σu_n converges by comparison with the convergent geometric progression Σk^n, $k < 1$.

EXERCISES 6(b)

1. Using the comparison test of Section 6.5 decide if the following series converge or diverge:

(a) $1 + \frac{1}{9} + \frac{1}{25} + \frac{1}{49} + \cdots$

(b) $\frac{1}{1.3} + \frac{1}{2.4} + \frac{1}{3.5} + \cdots$

(c) $\frac{1}{3.4} + \frac{2}{5.6} + \frac{3}{7.8} + \cdots$

(d) $\frac{1+2}{2^3} + \frac{1+2+3}{3^3} + \frac{1+2+3+4}{4^3}$

(e) $\Sigma\frac{n^{1/2}}{2n^2+1}$

(f) $\Sigma\frac{n+1}{3n^4-1}$

(g) $\Sigma\frac{10^n}{n!}$

(h) $\Sigma\frac{(n+2)(n+4)}{(n+1)(n+3)(n+5)}$

(i) $\sum \dfrac{2^n - 1}{5^n - 1}$ (j) $\sum \dfrac{1}{n} \sin \dfrac{x}{n}$

(k) $\sum\limits_{n=1}^{\infty} \dfrac{a^{n-1}}{1 + a^n}$ $(a > 0)$ (l) $\sum n^{(1/n)-2}$.

2. Prove that, if $u_n > 0$ and Σu_n is convergent, so also are

$$\Sigma u_n^2, \ \Sigma \frac{u_n}{n}, \ \Sigma \frac{u_n}{1 + u_n}, \ \Sigma \frac{u_n}{1 - u_n}, \ u_n \neq 1.$$

3. Given that u_n, c_n, d_n are all positive, Σc_n is convergent and Σd_n divergent, prove that

(a) if $\dfrac{u_n}{u_{n+1}} \geqslant \dfrac{c_n}{c_{n+1}}$, then Σu_n is convergent;

(b) if $\dfrac{u_n}{u_{n+1}} \leqslant \dfrac{d_n}{d_{n+1}}$, then Σu_n is divergent.

4. Test the following series for convergence:

(a) $\sum \dfrac{1}{n.3^n}$ (b) $\sum \dfrac{1}{n} \left(\dfrac{e}{\pi - 1}\right)^n$ (c) $\Sigma n^2 e^{an}$

(d) $1 + \dfrac{1}{2^2} + \dfrac{1}{3^2} + \dfrac{1}{4^{2/3}} + \dfrac{1}{5^2} + \dfrac{1}{6^2} + \dfrac{1}{7^2} + \dfrac{1}{8^2} + \dfrac{1}{9^{2/3}} + \cdots$,

that is, $u_n = n^{-2/3}$ when n is a perfect square, and $u_n = n^{-2}$ otherwise.

(e) $\Sigma 2^{-2n-(-1)^n}$ (f) $\Sigma n^{-3-(-1)^n}$

5. By grouping terms as in the example in Section 6.4, prove *Cauchy's condensation test*:

If $\Sigma f(n)$ is a series of positive monotonic decreasing terms, then it converges or diverges with $\sum\limits_{k=0}^{\infty} 2^k f(2^k)$.

6. Using the test in Question 5 above, test for convergence:

(a) $\sum \dfrac{1}{n(\log n)^\alpha}$ (b) $\sum \dfrac{1}{n^\alpha \log n}$

6.7 D'Alembert's ratio test

Theorem

(a) If $u_n > 0$ and there is a fixed number k and an integer n_0 such that

$$\frac{u_n}{u_{n+1}} \geqslant k > 1 \qquad \text{for } n \geqslant n_0,$$

then Σu_n converges.

(b) If $u_n > 0$ and $u_n/u_{n+1} < 1$ for $n \geqslant n_0$, where n_0 is an integer, then $u_n \nrightarrow 0$ as $n \to \infty$, and Σu_n is divergent.

(c) Limit form:

If $u_n > 0$ and $\lim_{n \to \infty} \dfrac{u_n}{u_{n+1}} = l$, then

if $l > 1$, Σu_n converges, and

if $l < 1$, Σu_n diverges.

Proof. (a) Since $u_n > 0$ we can write

$$u_{n+1} \leqslant \frac{1}{k} u_n, \qquad n \geqslant n_0.$$

Hence

$$u_n \leqslant \frac{1}{k} u_{n-1} \leqslant \frac{1}{k^2} u_{n-2} \leqslant \cdots \leqslant \frac{A}{k^n}, \qquad \text{where } A = u_{n_0} k^{-n_0}.$$

Thus Σu_n converges by comparison with the convergent geometric progression Σk^{-n}, $k > 1$.

(b) Since $u_n > 0$, the inequality may be written

$$u_{n+1} > u_n \qquad \text{for } n \geqslant n_0.$$

This implies

$$u_n > u_{n-1} > u_{n-2} > \cdots > u_{n_0}, \text{ and}$$

consequently

$$u_n \nrightarrow 0 \quad \text{as } n \to \infty.$$

Therefore Σu_n diverges.

To prove the limit form of the test we reduce it to cases (a) and (b) as in the proof of Cauchy's root test, Section 6.6.

Generally it is best to use the limit form of the test; it should be noted however, that when $l = 1$ the test is inconclusive.

It is important that a *fixed* k should exist in (a); for both $u_n = 1/n$ and $u_n = 1/n^2$ we have $u_n/u_{n+1} > 1$, and Σn^{-1} diverges and Σn^{-2} converges.

The ratio test and Raabe's test (Exercise 6(c), No. 3) are both particular cases of Kummer's test (Exercise 6(c), No. 2), a more comprehensive test.

6.8 Gauss's test

Theorem If $u_n > 0$ and if, for $n \geqslant n_0$,

$$\frac{u_n}{u_{n+1}} = 1 + \frac{\mu}{n} + \frac{A_n}{n^\lambda},$$

where A_n is bounded and $\lambda > 1$, then the series Σu_n is convergent if $\mu > 1$ and divergent if $\mu \leqslant 1$.

Proof. Notice that

$$\frac{u_n}{u_{n+1}} = 1 + \frac{\mu}{n} + \frac{A_n}{n^\lambda}, \qquad n \geqslant n_0$$

implies that u_n is eventually of constant sign (we take $u_n > 0$) and

$$n\left(\frac{u_n}{u_{n+1}} - 1\right) \to \mu \qquad \text{as } n \to \infty.$$

This is *Raabe's test* (See Exercises 6(c), No. 3); thus Σu_n converges if $\mu > 1$ and diverges if $\mu < 1$. The proof in the case $\mu = 1$ will be postponed until Exercise 8(c) No. 10.

Example 1 $u_n = 10^n/n$.

We have $u_n/u_{n+1} = (n+1)/10n \to \frac{1}{10}$ as $n \to \infty$. Hence Σu_n diverges by the ratio test.

Example 2 $u_n = x^n/n^2, x > 0$.

Here

$$\frac{u_n}{u_{n+1}} = \left(\frac{n+1}{n}\right)^2 \frac{1}{x} \to \frac{1}{x} \qquad \text{as } n \to \infty.$$

Then, by the ratio test, $\Sigma n^{-2}x^n$ converges for $0 < x < 1$ and diverges for $x > 1$. When $x = 1$,

$$\frac{u_n}{u_{n+1}} = \left(1 + \frac{1}{n}\right)^2 = 1 + \frac{2}{n} + \frac{1}{n^2}$$

and the convergence of Σu_n follows from Gauss's test.

Example 3 $u_n = x^n/n!, x > 0$.

Here

$$\frac{u_n}{u_{n+1}} = \frac{n+1}{x} \to \infty \qquad \text{as } n \to \infty,$$

and the series converges for $x > 0$.

Alternatively we can use the (a) form of the ratio test, for the infinite limit means that for any fixed $k > 1$

$$\frac{u_n}{u_{n+1}} > k > 1 \qquad \text{for } n > N.$$

Example 4 $$u_n = \frac{1.3.5 \cdots (2n-3)}{2.4.6 \cdots (2n)} x^n, \ x > 0.$$

Here

$$\frac{u_n}{u_{n+1}} = \left(\frac{2n+2}{2n-1}\right)\frac{1}{x} = \frac{1+1/n}{1-1/2n}\frac{1}{x} \to \frac{1}{x} \qquad \text{as } n \to \infty.$$

Therefore Σu_n converges for $0 < x < 1$ and diverges for $x > 1$. When $x = 1$,

$$\begin{aligned} \frac{u_n}{u_{n+1}} &= \left(1+\frac{1}{n}\right)\left(1-\frac{1}{2n}\right)^{-1} \\ &= \left(1+\frac{1}{n}\right)\left(1+\frac{1}{2n}+\frac{A_n}{n^2}\right), \qquad \text{where } A_n < \tfrac{1}{2}^{*} \\ &= 1+\frac{3/2}{n}+\frac{B_n}{n^2} \quad \left(\text{where } B_n = A_n + \tfrac{1}{2} + \frac{1}{n} < 2\right). \end{aligned}$$

Then Σu_n converges by Gauss's test.

EXERCISES 6(c)

1. Test the convergence or divergence of the series Σu_n for $x > 0$ when u_n is given by:

(a) nx^{n-1} (b) $\dfrac{x^n}{n(n+1)}$ (c) $\dfrac{x^n}{n!}$

(d) $\dfrac{1}{nx^n}$ (e) $\dfrac{n!}{x^n}$ (f) $n^2 x^n$

(g) $\dfrac{1.3.5 \cdots (2n-1)}{2.5.8 \cdots (3n+1)} x^n$ (h) $\dfrac{x^n}{n(x+n)}$

* By long division $(1-x)^{-1} = 1 + x + \dfrac{x^2}{1-x}$.

Then

$$\left(1-\frac{1}{2n}\right)^{-1} = 1 + \frac{1}{2n} + \frac{1}{4n^2-2n},$$

and

$$4n^2 - 2n > 4n^2 - 2n^2 = 2n^2, \ n > 1.$$

Hence $A_n < \frac{1}{2}$.

(i) $\dfrac{(n!)^2 x^{2n+1}}{(2n)!}$ (j) $\dfrac{(n!)^2 x^n}{(2n+1)!}$

2. Prove *Kummer's test*:
If $u_n > 0$, $\Sigma(d_n)^{-1}$ is a divergent series of positive terms and

$$\phi(n) = d_n \frac{u_n}{u_{n+1}} - d_{n+1},$$

then if $\lim_{n\to\infty} \phi(n) = l$, Σu_n converges if $l > 0$ and diverges if $l < 0$.

3. By taking $d_n = n$ in Kummer's test, deduce *Raabe's test*:

$$\text{If } u_n > 0 \quad \text{and} \quad \lim_{n\to\infty} n\left(\frac{u_n}{u_{n+1}} - 1\right) = l,$$

Σu_n converges if $l > 1$ and diverges if $l < 1$.

4. Deduce the ratio test from Kummer's test.

5. Given

$$a_n = \frac{2^2}{1.3}\frac{4^2}{3.5}\frac{6^2}{5.7}\cdots\frac{(2n)^2}{(2n-1)(2n+1)},$$

show that

$$\frac{a_n - a_{n-1}}{a_{n+1} - a_n} = 1 + \frac{2}{n} + \frac{3}{4n^2}.$$

Deduce that a_n tends to a finite limit as $n \to \infty$.

6. If u_n is positive and decreasing and $\nu = [\frac{1}{2}n]$, show that $\sum_{r=\nu+1}^{n} u_r \geqslant \frac{1}{2}nu_n$.
Hence prove that if Σu_n is a convergent series of positive decreasing terms, then $\lim_{n\to\infty} nu_n = 0$ (*Pringsheim's test*).

7. Use Pringsheim's test to show that

$$\sum n/(n^2 + a^2) \quad \text{and} \quad \sum n^{-1-1/n}$$

are both divergent.

8. (a) Show that $\lim_{n\to\infty} nu_n = 0$ is necessary but not sufficient for the convergence of Σu_n, when u_n is positive and decreasing.

(b) By considering Σu_n, where $u_n = n^{-1}$ if n is a perfect square and $u_n = n^{-2}$ otherwise, show that in Pringsheim's test u_n must be monotonic.

CHAPTER 7

Absolute Convergence and Power Series

7.1 Alternating series

In most of the preceding chapter we were restricted to series of positive terms. In this chapter we drop that restriction and consider series whose terms are any real numbers. The simplest series of this type is one whose terms are alternately positive and negative; such a series is called an *alternating series*. It is convenient to write such a series $\Sigma(-1)^{n-1} u_n$ where $u_n > 0$.

Theorem (Alternating series test) The alternating series $\Sigma(-1)^{n-1} u_n$, $u_n > 0$ is convergent if u_n is monotonic decreasing and null.

Proof. Let $s_n = \sum_{r=1}^{n} (-1)^{r-1} u_r$; then

$$
\begin{aligned}
s_{n+p} - s_n &= \sum_{r=n+1}^{n+p} (-1)^{r-1} u_r \\
&= (-1)^n \{u_{n+1} - u_{n+2} + u_{n+3} - \cdots + (-1)^p u_{n+p}\} \\
&= (-1)^n \{u_{n+1} - (u_{n+2} - u_{n+3}) - \cdots - (u_{n+p-2} - u_{n+p-1}) - u_{n+p}\} \\
&\quad \text{if } p \text{ is even;} \\
&= (-1)^n \{u_{n+1} - (u_{n+2} - u_{n+3}) - \cdots - (u_{n+p-1} - u_{n+p})\} \\
&\quad \text{if } p \text{ is odd.}
\end{aligned}
$$

Since u_n is monotonic decreasing, $u_r - u_{r+1} \geqslant 0$ for $r = 1, 2, 3, \ldots$ and $|s_{n+p} - s_n| \leqslant u_{n+1}$. Since (u_n) is null,

$$\forall \epsilon > 0; \exists m.\ u_n < \epsilon \qquad \text{for } n \geqslant m.$$

Therefore $|s_{n+p} - s_n| < \epsilon$ for $n \geqslant m$, $p = 1, 2, 3, \ldots$ and $\Sigma(-1)^{n-1} u_n$ converges by the general principle of convergence.

Corollary If $\Sigma(-1)^{n-1} u_n$, $u_n > 0$ converges to the sum s, then s lies between s_n and s_{n+1} for all n.

We have seen that

$$|s_{n+p} - s_n| \leqslant u_{n+1}.$$

Let $p \to \infty$ holding n fixed; then

$$|s - s_n| \leqslant u_{n+1}.$$

In other words, the error in taking the sum to be s_n is not greater numerically than the modulus of the first term omitted.

Example 1 $1 - \frac{1}{2} + \frac{1}{3} - \frac{1}{4} + \cdots$ converges by the alternating series test, for $(1/n)$ is a decreasing null sequence.

Example 2 If u_n is not a decreasing function of n, the test cannot be used. Thus in the convergent series

$$\sum_{n=0}^{\infty} (-1)^n u_n = 1 - \frac{1}{2} + \frac{1}{2^4} - \frac{1}{2^3} + \frac{1}{2^8} - \frac{1}{2^5} + \cdots + \frac{1}{2^{4r}} - \frac{1}{2^{2r+1}} + \cdots,$$

u_n is not monotonic, for

$$\frac{u_n}{u_{n+1}} = \begin{cases} 2^{-2r+1} < 1 & \text{when } n = 2r, \\ 2^{2r+3} > 1 & \text{when } n = 2r+1, \end{cases}$$

$$r = 0, 1, 2, \ldots.$$

On the other hand, the series $1 - \frac{1}{2} + \frac{1}{2} - \frac{1}{4} + \frac{1}{3} - \frac{1}{8} + \frac{1}{4} - \cdots$, which is formed by subtracting $\frac{1}{2} + \frac{1}{4} + \frac{1}{8} + \cdots$ from $1 + \frac{1}{2} + \frac{1}{3} + \cdots$ is divergent.

Example 3

$$u_n = \frac{1}{n^{1/2} + (-1)^{n-1}}.$$

We have

$$(-1)^n u_n = (-1)^n \left\{ \frac{n^{1/2} - (-1)^{n-1}}{n-1} \right\} = (-1)^n a_n + b_n.$$

$\Sigma(-1)^n a_n$ is convergent by the alternating series test, and Σb_n is divergent (the harmonic series). Therefore $\Sigma(-1)^n u_n$ is divergent.

Example 4 If $u_n > 0$ and

$$\frac{u_n}{u_{n+1}} = 1 + \frac{\mu}{n} + \frac{A_n}{n^\lambda}, n \geqslant n_0,$$

where A_n is bounded and $\lambda > 1$, then if $\mu > 0$, u_n is decreasing and null.

We can write

$$\frac{u_n}{u_{n+1}} = 1 + \frac{\mu_n}{n},$$

where $\mu_n \to \mu$ as $n \to \infty$. Then $\mu_n > \frac{1}{2}\mu > 0$ for $n \geqslant m$, and

$$\frac{u_n}{u_{n+1}} > 1 + \frac{\frac{1}{2}\mu}{n} > 1, \qquad n \geqslant m,$$

so that u_n is decreasing. Also,

$$\begin{aligned} \frac{u_m}{u_n} &= \frac{u_m}{u_{m+1}} \frac{u_{m+1}}{u_{m+2}} \frac{u_{m+2}}{u_{m+3}} \cdots \frac{u_{n-1}}{u_n} \qquad \text{(where } m \text{ is fixed)} \\ &> \left(1 + \frac{\frac{1}{2}\mu}{m}\right)\left(1 + \frac{\frac{1}{2}\mu}{m+1}\right) \cdots \left(1 + \frac{\frac{1}{2}\mu}{n-1}\right) \\ &> 1 + \tfrac{1}{2}\mu \sum_{r=m}^{n-1} \frac{1}{r}. \end{aligned}$$

Now the harmonic series is divergent, so $\sum_{r=m}^{n-1} r^{-1}$ is unbounded as $n \to \infty$, consequently $u_n \to 0$ as $n \to \infty$.

7.2 Absolute convergence

A series Σu_n is said to be *absolutely convergent* if the series of moduli $\Sigma |u_n|$ is convergent. This definition implies that *an absolutely convergent series is convergent.*

For if $\Sigma |u_n|$ is convergent, then by the general principle of convergence,

$$\forall \epsilon > 0; \exists m. \sum_{n=m+1}^{m+p} |u_n| < \epsilon \qquad \text{for } p = 1, 2, 3, \ldots.$$

Since

$$\left| \sum_{n=m+1}^{m+p} u_n \right| \leqslant \sum_{n=m+1}^{m+p} |u_n|,$$

it follows that Σu_n also converges.

The converse of this theorem is false. A counter-example justifying this statement is the alternating series $1 - \frac{1}{2} + \frac{1}{3} - \frac{1}{4} + \cdots$ whose convergence was established in the previous section. Here the series of moduli is the harmonic series which is divergent. A series which is convergent but not absolutely convergent is said to be *conditionally convergent*.

Since $|u_n|$ is non-negative, we may use the tests for positive series obtained in Chapter 6 to test the behaviour of $\Sigma |u_n|$. It is suggested that these tests are applied in the following order:

1. If $\lim_{n\to\infty} u_n \neq 0$, then $\lim_{n\to\infty} |u_n| \neq 0$ also and both Σu_n and $\Sigma |u_n|$ diverge.
2. Use a comparison test if $|u_n|$ is a rational function of n which can be compared with $\Sigma n^{-\alpha}$, with a suitable value of α.
3. D'Alembert's ratio test; this is a particularly useful test because the divergence part implies $|u_n| \nrightarrow 0$ as $n \to \infty$ and so establishes the divergence of Σu_n.
4. When the ratio test breaks down because

$$\lim_{n\to\infty} \frac{|u_n|}{|u_{n+1}|} = 1,$$

try Gauss's test.

It should be remembered that there is no universal test which will work for all series.

Example 1 $\Sigma(-1)^{n-1} x^n/n$.

Here

$$u_n = \frac{(-1)^{n-1} x^n}{n}$$

and

$$\frac{|u_n|}{|u_{n+1}|} = \frac{n+1}{n}\frac{1}{|x|} \to \frac{1}{|x|} \qquad \text{as } n \to \infty.$$

Then $\Sigma(-1)^{n-1} x^n/n$ is absolutely convergent for $|x| < 1$, and divergent for $|x| > 1$.

When $x = -1$, $u_n = 1/n$ and Σu_n is divergent (harmonic series); when $x = 1$, Σu_n converges by the alternating series test.

Example 2 $u_n = x^n/n^3$.

Using the ratio test

$$\frac{|u_n|}{|u_{n+1}|} = \left(\frac{n+1}{n}\right)^3 \frac{1}{|x|} \to \frac{1}{|x|} \qquad \text{as } n \to \infty.$$

Then $\Sigma x^n/n^3$ is absolutely convergent for $|x| < 1$ and divergent for $|x| > 1$.

When $|x| = 1$, $\Sigma|u_n| = \Sigma n^{-3}$, which is convergent, so that Σu_n is absolutely convergent for $|x| \leqslant 1$.

Example 3 $u_n = x^n/n!$.

The ratio test gives

$$\frac{|u_n|}{|u_{n+1}|} = \frac{n+1}{|x|} \to \infty \qquad \text{as } n \to \infty.$$

Thus $\Sigma x^n/n!$ is absolutely convergent for all x. (See Example 3, Section 6.8.)

Example 4 $u_n = \binom{m}{n} x^n$, where

$$\binom{m}{n} = \frac{m(m-1)(m-2)\cdots(m-n+1)}{n!}.$$

Using the ratio test we have

$$\frac{|u_n|}{|u_{n+1}|} = \frac{n+1}{|n-m|}\frac{1}{|x|} \to \frac{1}{|x|} \qquad \text{as } n \to \infty.$$

Thus Σu_n is absolutely convergent for $|x| < 1$ and divergent for $|x| > 1$ for all values of m. When $|x| = 1$,

$$\frac{|u_n|}{|u_{n+1}|} = \frac{1+1/n}{1-m/n} = 1 + \frac{m+1}{n} + \frac{A_n}{n^2}$$

where A_n is bounded.*

Thus, when $|x| = 1$, Σu_n is absolutely convergent if $m > 0$. When $-1 < m < 0$, $|u_n|$ is monotonic decreasing and $|u_n| \to 0$ as $n \to \infty$ (Example 4, Section 7.1). Also for this range of m, $\binom{m}{n}$ alternates in sign; therefore, by the alternating series test, Σu_n is convergent at $x = 1$ when $m > -1$.

EXERCISES 7(a)

1. Test for convergence:

(a) $1 - \frac{1}{3} + \frac{1}{5} - \frac{1}{7} + \cdots$ (b) $2 - \frac{3}{2} + \frac{4}{3} - \frac{5}{4} + \cdots$

* By long division

$$\left(1+\frac{1}{n}\right)\left(1-\frac{m}{n}\right)^{-1} = 1 + \frac{m+1}{n} + \frac{m(m+1)}{n^2}\frac{1}{1-m/n}$$

$1 - m/n > \frac{1}{2}$ when $n > 2m$. Hence $A_n < 2m(m-1)$.

(c) $1-\frac{3}{4}+\frac{5}{8}-\frac{7}{12}+\cdots$ (d) $1-\frac{3}{4}+\frac{5}{9}-\frac{7}{16}+\frac{9}{25}-\cdots$

(e) $1-\dfrac{1}{2!}+\dfrac{1}{3!}-\dfrac{1}{4!}+\cdots$ (f) $\sum \dfrac{(-1)^n}{\log n}$

(g) $\sum \dfrac{(-1)^n(n+1)}{n^2+5n+6}$ (h) $\sum \dfrac{(-1)^n(2n+1)}{(n+1)(n+2)}$

(i) $\sum \dfrac{(-10)^n}{n!}$ (j) $\sum \dfrac{(-1)^n n\pi^n}{e^{2n}+1}$

(k) $\sum (-1)^n\left\{\dfrac{1}{n^{1/2}}+\dfrac{(-1)^{n-1}}{n}\right\}$ (l) $\sum \dfrac{(-1)^n}{n+2+(-1)^n}$

2. Test $\sum\limits_{r=1}^{\infty}(-1)^{r-1}u_r$ for convergence when

(a) $u_{2r}=3^{-2r},\ u_{2r-1}=2^{-2r-1}$,

(b) $u_{2r}=4^{-2r},\ u_{2r-1}=3^{-2r-1}$,

(c) $u_r=r^{-1/2}$, r even, $u_r=r^{-3/2}$, r odd.

3. For the series in Question 2(a) above show that $|s-s_n|>|u_{n+1}|$.

4. Prove that, if Σu_n is conditionally convergent then the series of positive terms and the series of negative terms taken separately are both divergent.

5. Prove that a real series is absolutely convergent if the series of positive terms and the series of negative terms are both convergent.

6. By considering the series

$$\sum_{n=1}^{\infty}\frac{(-1)^n}{\sqrt{n}} \quad\text{and}\quad \sum_{n=1}^{\infty}\frac{(-1)^n}{\sqrt{n}+(-1)^n\log n}$$

show that one of the series Σu_n, Σv_n may be convergent and the other divergent even though $u_n/v_n\to 1$ as $n\to\infty$, when all the terms are not of the same sign.

7. Discuss the convergence and absolute convergence of Σu_n when x is real and u_n is given by:

(a) nx^{n-1} (b) $\dfrac{(-x)^n}{2n-1}$ (c) $\dfrac{x^{4n}}{n^4}$

(d) $\left(\dfrac{n+1}{n}\right)x^n$ (e) $(-1)^{n-1}\dfrac{x^n}{n^2}$ (f) $\dfrac{\cos nx}{n\sqrt{n}}$

(g) $(-1)^n \cos \frac{x}{n}$ (h) $\frac{x^n}{(\log n)^2}$

(i) $\frac{\text{sh}\, nx}{\log n}$ (j) $\left(\frac{n^2-1}{n^2+1}\right) x^n$

(k) $\frac{(n!)^2 x^{2n}}{(2n)!}$ (l) $\left(\frac{(2n!)}{2^{2n}(n!)^2}\right)^\alpha x^n$

(m) $\frac{2^n n! x^n}{1.5.9 \cdots (4n+1)}$ (n) $(-1)^n n! \left(\frac{x}{n}\right)^n$

(o) $\frac{x^n}{\sqrt{n}} \log \left(\frac{2n+1}{n}\right)$.

8. If Σa_n is absolutely convergent and (b_n) is a bounded sequence, prove that $\Sigma a_n b_n$ is absolutely convergent.

7.3 Further tests for convergence

Let (a_n), (b_n) be any two sequences of real numbers and let $A_n = \sum_{r=1}^{n} a_r$; then for positive integers $n, k \geqslant 1$,

$$\sum_{r=n+1}^{n+k} a_r b_r = \sum_{r=n+1}^{n+k} A_r(b_r - b_{r+1}) - A_n b_{n+1} + A_{n+k} b_{n+k+1}.$$

This algebraic identity is known as *Abel's partial summation* formula and follows by summing

$$\begin{aligned} a_r b_r &= (A_r - A_{r-1}) b_r \\ &= A_r(b_r - b_{r+1}) - A_{r-1} b_r + A_r b_{r+1}, \end{aligned}$$

for $r = n+1, n+2, \ldots, n+k$.

Theorem (*Abel's test*) If Σa_n is convergent and (b_n) is a positive decreasing sequence, then $\Sigma a_n b_n$ converges.

Proof. Suppose $b_n \to b \geqslant 0$ as $n \to \infty$. If $b > 0$, let $c_n = b_n - b$, then (c_n) is a decreasing null sequence. Hence

$$\forall \epsilon > 0; \exists N_1 \; . \; c_n < q\epsilon \qquad \text{for } n \geqslant N_1,$$

where q is positive. Since Σa_n is convergent, its partial sums are bounded,

$$|A_n| \leqslant M \qquad \text{for all } n,$$

and, by the general principle of convergence

$$|A_{n+p} - A_n| < q'\epsilon, \qquad n \geqslant N_2, \quad p = 1, 2, 3, \ldots$$

where q' is positive.

Writing $b_n = c_n + b$ in Abel's partial summation formula, we have

$$\begin{aligned}\left|\sum_{r=n+1}^{n+k} a_r b_r\right| &= \left|\sum_{r=n+1}^{n+k} A_r(c_r - c_{r+1}) - A_n c_{n+1} + A_{n+k}c_{n+k+1} \right.\\ &\qquad \left. + b(A_{n+k} - A_n)\right| \\ &\leqslant M \sum_{r=n+1}^{n+k} (c_r - c_{r+1}) + M(c_{n+1} + c_{n+k+1}) + b|A_{n+k} - A_n| \\ &\leqslant 2Mc_{n+1} + b|A_{n+k} - A_n|, \\ &< 2Mq\epsilon + bq'\epsilon \qquad \text{for } n \geqslant N = \max(N_1, N_2) \text{ and} \\ &\qquad k = 1, 2, 3, \ldots .\end{aligned}$$

Now choose $q < 1/4M$, $q' < 1/2b$ and we have

$$\left|\sum_{r=n+1}^{n+k} a_r b_r\right| < \epsilon \qquad \text{for } n \geqslant N, \quad k = 1, 2, 3, \ldots .$$

Hence $\Sigma a_n b_n$ converges by the general principle of convergence.

Theorem (*Dirichlet's test*) If A_n is bounded and (b_n) is a positive, decreasing null sequence, the $\Sigma a_n b_n$ is convergent.

Proof. Let $|A_n| \leqslant M$ for all n. Since (b_n) is null,

$$b_n < \frac{\epsilon}{2M} \qquad \text{for } n \geqslant N.$$

Then, using Abel's partial summation formula. we have

$$\begin{aligned}\left|\sum_{r=n+1}^{n+k} a_r b_r\right| &\leqslant M\left\{\sum_{r=n+1}^{n+k} (b_r - b_{r+1}) + b_{n+1} + b_{n+k+1}\right\} \\ &\leqslant 2Mb_{n+1} \\ &< \epsilon \qquad \text{for } n \geqslant N.\end{aligned}$$

The convergence of $\Sigma a_n b_n$ follows from the general principle of convergence.

Example 1 If Σa_n is convergent then so also are $\Sigma a_n x^n$ and $\Sigma a_n n^{-x}$ for $0 \leqslant x \leqslant 1$ by Abel's test.

Example 2 $\Sigma n^{-1} \cos nx$ converges for $0 < x < 2\pi$.

We first show that $\Sigma \cos nx$ has bounded partial sums. Let

$$C_n = \sum_{r=1}^{n} \cos rx, \quad S_n = \sum_{r=1}^{n} \sin rx;$$

then

$$C_n + iS_n = \sum_{r=1}^{n} e^{irx} = \frac{e^{ix}(1-e^{inx})}{1-e^{ix}}, \qquad x \neq 2k\pi, \quad k = 0, \pm 1, \pm 2, \ldots$$

$$= \frac{e^{i\frac{1}{2}x}(1-e^{inx})}{e^{-i\frac{1}{2}x} - e^{i\frac{1}{2}x}}$$

$$= \frac{e^{i\frac{1}{2}x} - e^{i(n+1/2)x}}{-2i \sin \frac{1}{2}x}$$

Then equating real and imaginary parts,

$$C_n = \frac{\sin (n+\frac{1}{2})x - \sin \frac{1}{2}x}{-2i \sin \frac{1}{2}x}, \quad S_n = \frac{\cos \frac{1}{2}x - \cos (n+\frac{1}{2})x}{2 \sin \frac{1}{2}x}$$

provided $x \neq 2k\pi$, $k = 0, \pm 1, \pm 2, \ldots$.

Hence $|C_n| \leqslant \operatorname{cosec} \frac{1}{2}x$, $0 < x < 2\pi$. Since $b_n = n^{-1}$ is decreasing and null, $\Sigma n^{-1} \cos nx$ converges for $0 < x < 2\pi$ by Dirichlet's test.

7.4 Series of complex terms

If Σz_n is a series of complex numbers, $z_n = x_n + iy_n$, where x_n and y_n are real, then Σz_n converges if and only if both Σx_n and Σy_n converge; if $\Sigma x_n = \xi$ and $\Sigma y_n = \eta$, then $\Sigma z_n = \xi + i\eta$. In terms of partial sums

$$Z_n = \sum_{r=1}^{n} (x_r + iy_r) = \sum_{r=1}^{n} x_r + i \sum_{r=1}^{n} y_r$$

$$= X_n + iY_n,$$

then $\lim_{n\to\infty} Z_n$ exists if and only if $\lim_{n\to\infty} X_n$ and $\lim_{n\to\infty} Y_n$ both exist. The series Σz_n is *absolutely convergent* if $\Sigma |z_n|$ is convergent. Now, when $z_n = x_n + iy_n$,

$$|x_n|, |y_n| \leqslant |z_n| \leqslant |x_n| + |y_n|.$$

Hence, from the left-hand inequalities, when Σz_n is absolutely convergent then so also are Σx_n, Σy_n. Since x_n and y_n are real, this implies that Σx_n and Σy_n are convergent, and consequently $\Sigma(x_n + iy_n) = \Sigma z_n$ is convergent. That is, *absolute convergence implies convergence for complex series.*

Also, the above inequalities show that Σz_n is absolutely convergent if and only if both Σx_n and Σy_n are absolutely convergent.

Since Σx_n, Σy_n and $\Sigma |z_n|$ are all real, then the tests developed earlier may all be applied to these series. However, since there is no order in complex numbers, tests involving inequalities or monotonic sequences only apply when the quantities involved are real.

Example 1 $\Sigma z^n/n^3$ is absolutely convergent for $|z| \leqslant 1$.

For if $z_n = z^n/n^3$,

$$\frac{|z_n|}{|z_{n+1}|} = \left(\frac{n+1}{n}\right)^3 \frac{1}{|z|} \to \frac{1}{|z|} \qquad \text{as } n \to \infty,$$

and $\Sigma n^{-3} z^n$ is absolutely convergent for $|z| < 1$ by the ratio test; also the series is divergent for $|z| > 1$. When $|z| = 1$, $\Sigma |z_n| = \Sigma n^{-3}$, which is convergent. Therefore $\Sigma z^n/n^3$ is absolutely convergent for $|z| \leqslant 1$ and divergent for $|z| > 1$.

Example 2 $\Sigma z^{4n}/n$ is absolutely convergent for $|z| < 1$.

For if $z_n = z^{4n}/n$,

$$\frac{|z_n|}{|z_{n+1}|} = \left(\frac{n+1}{n}\right) \frac{1}{|z|^4} \to \frac{1}{|z|^4} \qquad \text{as } n \to \infty,$$

and Σz_n is absolutely convergent for $|z| < 1$, divergent for $|z| > 1$ by the ratio test. When

$$|z| = 1, z = e^{i\theta}$$

and

$$\Sigma \frac{z^{4n}}{n} = \Sigma\, n^{-1} \cos 4n\theta + i \Sigma\, n^{-1} \sin 4n\theta.$$

Now, using the result of Example 2, Section 7.3, we know that $\sum_{r=1}^{n} \cos 4r\theta$ and $\sum_{r=1}^{n} \sin 4r\theta$ are bounded provided $\theta \neq 0, \pm\frac{1}{2}\pi, \pi$. Hence by Dirichlet's test $\Sigma z^{4n}/n$ converges for $|z| = 1$ except when $z = \pm 1, \pm i$.

EXERCISES 7(b)

1. Discuss, for $0 \leqslant x \leqslant 2\pi$, the convergence of the series:

(a) $\Sigma \dfrac{\cos nx}{\sqrt{n}}$ (b) $\Sigma \dfrac{\sin nx}{n}$ (c) $\Sigma \dfrac{\sin nx}{n^2}$.

2. Show that $\Sigma n^{-\alpha} \cos nx$ is absolutely convergent for all real x if $\alpha > 1$, and is convergent for $0 < x < 2\pi$ if $0 < \alpha < 1$.

3. Discuss the convergence of

$$\text{(a) } \sum \frac{(-1)^n}{n} \cos \frac{x}{n} \qquad \text{(b) } \sum \frac{n}{n^2+1} \sin nx.$$

4. Prove that, if both Σa_n and $\Sigma |b_n - b_{n+1}|$ converges, so also does $\Sigma a_n b_n$.

Show that if the series $c_1 + 2c_2 + 3c_3 + \cdots$ converges, then the series $c_{k+1} + 2c_{k+2} + 3c_{k+3} + \cdots$ also converges for each positive integer k. Show that if its sum is denoted by $S(k)$, then $S(k) \to 0$ as $k \to \infty$.

5. Prove that if $u_1, u_2, u_3, \ldots, u_n$ is a set of real numbers such that

$$A < \sum_{r=1}^{m} u_r < B, \qquad m = 1, 2, \ldots, n,$$

where A and B are fixed numbers, and $a_1, a_2, \ldots, a_n$ is a set of monotonic decreasing numbers, then

$$a_1 A < \sum_{r=1}^{m} a_r u_r < a_1 B, \qquad m = 1, 2, \ldots, n.$$

Show that if x is not a multiple of π and the principal value of the inverse tangent is taken,

$$\sum_{r=1}^{m} (\sin rx) \tan^{-1} r$$

oscillates between two bounds differing at most by $\frac{1}{2}\pi \operatorname{cosec} \frac{1}{2}x$.

6. Discuss the convergence of the following series for complex values of z:

$$\text{(a) } \sum nz^n \qquad \text{(b) } \sum \frac{z^n}{n} \qquad \text{(c) } \sum \frac{z^n}{n^2}$$

$$\text{(d) } \sum \frac{z^{2n}}{\sqrt{n}} \qquad \text{(e) } \sum \frac{nz^n}{(n+1)(n+2)}$$

7. Given that $z_1, z_2, z_3, \ldots$ are real or complex numbers and $\Sigma |z_n|$ converges, prove that Σz_n converges.

Prove that if $|\arg z_n| < \frac{1}{2}\pi - \delta, \delta > 0$ $(n = 1, 2, 3, \ldots)$, then the convergence of Σz_n implies the convergence of $\Sigma |z_n|$.

7.5 Rearrangement of series

If (k_n) is a sequence whose elements are the positive integers in some order, then Σu_{k_n} is called a *rearrangement* of the series Σu_n. That is, a rearrangement is a series formed by taking the terms of a given series in a different order, so that each term of the original series is included once and once only. For example, $1 + \frac{1}{3} - \frac{1}{2} + \frac{1}{5} + \frac{1}{7} - \frac{1}{4} + \cdots$ is a rearrangement of the series $1 - \frac{1}{2} + \frac{1}{3} - \frac{1}{4} + \frac{1}{5} - \frac{1}{6} + \cdots$; it was seen in Section 6.2 (Property 6) that the two series have different sums. We are now going to show that *if Σu_n is absolutely convergent, then a rearrangement of the series does not change its sum.*

First we prove the result for positive series. Let $u_n \geqslant 0$ and suppose Σu_n converges to a sum s, then

$$\forall \epsilon > 0; \exists N.\ |s - s_n| < \epsilon \qquad \text{for } n \geqslant N.$$

Hence

$$s_n > s - \epsilon, \qquad n \geqslant N.$$

Now let Σv_n be any rearrangement of Σu_n, and let $\sigma_n = \sum_{r=1}^{n} v_r$. Since Σv_n consists of the terms of Σu_n taken in a different order, there is an integer m such that σ_m contains *all* the terms of s_N. Then

$$\sigma_m \geqslant s_N > s - \epsilon.$$

But, since Σv_n contains only the terms of Σu_n and no others,

$$\sigma_n \leqslant s \qquad \text{for all } n.$$

Therefore

$$s - \epsilon < \sigma_n \leqslant s \qquad \text{for } n \geqslant m;$$

that is Σv_n converges to s.

Now suppose Σu_n is absolutely convergent. Let

$$p_n = \tfrac{1}{2}(|u_n| + u_n) \quad \text{and} \quad q_n = \tfrac{1}{2}(|u_n| - u_n),$$

then $p_n, -q_n$ represent the positive and negative terms of the series Σu_n; also $u_n = p_n - q_n$ and $|u_n| = p_n + q_n$. Then $p_n \geqslant 0, q_n \geqslant 0$ and $p_n \leqslant |u_n|, q_n \leqslant |u_n|$. Therefore Σp_n, Σq_n both converge by comparison with the convergent series $\Sigma |u_n|$, and the result follows from that for positive series.

Example 1 $1 - (\frac{1}{2})^2 + (\frac{1}{3})^2 - (\frac{1}{4})^2 + \cdots$ and the rearrangement $1 + (\frac{1}{3})^2 - (\frac{1}{2})^2 + (\frac{1}{5})^2 + (\frac{1}{7})^2 - (\frac{1}{4})^2 + \cdots$ both have the same sum as the series are absolutely convergent.

Here

$$\sigma_{3n} = 1 + \frac{1}{3^2} - \frac{1}{2^2} + \cdots + \frac{1}{(4n-3)^2} + \frac{1}{(4n-1)^2} - \frac{1}{(2n)^2}$$

$$= s_{2n} + \left\{\frac{1}{(2n+1)^2} + \frac{1}{(2n+3)^2} + \cdots + \frac{1}{(4n-1)^2}\right\}.$$

Then

$$\sigma_{3n} - s_{2n} < \frac{n}{(2n+1)^2}$$

$$\to 0 \qquad \text{as } n \to \infty.$$

Example 2 In contrast, notice that when the series $1 - \frac{1}{2} + \frac{1}{3} - \frac{1}{4} + \cdots$ and its rearrangement $1 + \frac{1}{3} - \frac{1}{2} + \frac{1}{5} + \frac{1}{7} - \frac{1}{4} + \cdots$ are considered, then

$$\sigma_{3n} - s_{2n} = \frac{1}{2n+1} + \frac{1}{2n+3} + \cdots + \frac{1}{4n-1},$$

and

$$\frac{n}{4n-1} < \sigma_{3n} - s_{2n} < \frac{n}{2n+1}.$$

That is $\sigma_{3n} - s_{2n}$ lies between $\frac{1}{4}$ and $\frac{1}{2}$ and so cannot tend to zero as n tends to infinity. Consequently the rearrangement has altered the sum. (See Exercise 9(f) No. 4(c).) There is a theorem due to Riemann which asserts that if Σu_n is conditionally convergent, then a suitable rearrangement makes the resulting series oscillate between two pre-assigned bounds. That is the rearranged series can be made to converge to a given sum or to diverge.

7.6 Multiplication of series

Theorem If $\sum_{n=0}^{\infty} u_n = U$ and $\sum_{n=0}^{\infty} v_n = V$ are absolutely convergent series, then the Cauchy product $\sum_{n=0}^{\infty} p_n$, where $p_n = \sum_{r=0}^{n} u_r v_{n-r}$, is absolutely convergent to UV.

Proof. The product series is

$$p_n = u_0 v_0 + (u_0 v_1 + u_1 v_0) + (u_0 v_2 + u_1 v_1 + u_2 v_0) + \cdots;$$

Let $\sum_{n=0}^{\infty} d_n$ be the series obtained by removing the brackets in Σp_n. This can be expressed as a double array

$$\begin{array}{ccccc} u_0 v_0 & u_0 v_1 & u_0 v_2 & u_0 v_3 & \cdots \\ u_1 v_0 & u_1 v_1 & u_1 v_2 & u_1 v_3 & \cdots \\ u_2 v_0 & u_2 v_1 & u_2 v_2 & u_2 v_3 & \cdots \\ u_3 v_0 & u_3 v_1 & u_3 v_2 & u_3 v_3 & \cdots \\ \cdot & \cdot & \cdot & \cdot & \cdot \end{array}$$

Let the partial sums be represented by

$$U_n = \sum_{r=0}^{n} u_r, \quad V_n = \sum_{r=0}^{n} v_r, \quad P_n = \sum_{r=0}^{n} p_r, \quad D_n = \sum_{r=0}^{n} d_r.$$

Then D_n is the sum of the first $n+1$ diagonals; $U_n V_n$ is the sum of the terms in the $(n+1)$th square formed by the terms common to the first $n+1$ rows and the first $n+1$ columns. Clearly the $(n+1)$th square contains all the terms of D_n together with some additional ones.

First let $u_n \geqslant 0$ and $v_n \geqslant 0$; then $D_n \leqslant U_n V_n \leqslant UV$ for every n. Also D_n, U_n and V_n are all monotonic increasing; therefore Σd_n converges and $\lim_{n\to\infty} D_n = D \leqslant UV$. Again, $U_n V_n \leqslant D_{2n} \leqslant D$, since Σd_n converges to D. Also (U_n) converges to U and (V_n) converges to V; therefore $UV \leqslant D$. Combining the two results shows that $D = UV$. Hence (p_n), which is a subsequence of (D_n), also converges to UV.

If Σu_n and Σv_n are absolutely convergent, form the positive series Σa_n, Σb_n, $\Sigma a_n'$ and $\Sigma b_n'$ by writing (as in Section 7.5)

$$a_n = \tfrac{1}{2}(|u_n| + u_n), \quad b_n = \tfrac{1}{2}(|u_n| - u_n), \quad a_n' = \tfrac{1}{2}(|v_n| + v_n),$$
$$b_n' = \tfrac{1}{2}(|v_n| - v_n).$$

Then the above theorem can be applied to these series and since the product series is a linear combination of these positive series the result follows.

It can be shown that if Σu_n, Σv_n are both convergent, but not absolutely convergent, then Σp_n may diverge (see Exercises 7(c) No. 5), but if Σp_n does converge then its sum is UV (see Example 2, Section 7.10).

EXERCISES 7(c)

1. A convergent alternating series $s = u_1 - u_2 + u_3 - u_4 + \cdots$ is rearranged to $u_1 - u_2 - u_4 + u_3 - \cdots$ and the partial sums of the original series and the rearranged series are denoted by s_n and σ_n respectively. Show that

$$s_{2n} - \sigma_{3n} = \sum_{r=n+1}^{2n} u_{2r}.$$

Show also that (a) if $u_n = n^{-1/2}$, then the rearranged series diverges to $-\infty$; and (b) if $u_n = n^{-3/2}$, then the rearranged series converges to s.

2. Show that

$$\sum_{n=1}^{\infty} \frac{6n-3}{n(4n-3)(4n-1)} = \log 2.$$

3. Show that if Σa_n converges and $|x| < 1$, then

$$(1-x)^{-1} \sum_{n=0}^{\infty} a_n x^n = \sum_{n=0}^{\infty} s_n x^n,$$

where $s_n = a_0 + a_1 + a_2 + \cdots + a_n$.

4. Show that if $E(x) = \sum_{n=0}^{\infty} x^n/n!$, then $E(x)E(y) = E(x+y)$.

5. Show that if $a_n = (-1)^n/\sqrt{(n+1)}$ and $(\sum_{n=0}^{\infty} a_n)^2 = \sum_{n=0}^{\infty} p_n$, then

$$p_n = (-1)^n \sum_{r=0}^{n} \frac{1}{\sqrt{\{(r+1)(n+1-r)\}}}.$$

Show also that Σa_n is convergent but Σp_n is divergent.

7.7 Power series

A power series is a series of the form $\sum_{n=0}^{\infty} a_n x^n$ or $\sum_{n=0}^{\infty} a_n(x-a)^n$, where a_n, a, x are real or complex numbers and a, a_n are independent of x. For simplicity we shall concentrate on power series of the form $\Sigma a_n x^n$. Clearly every such series is convergent at $x = 0$.

Theorem If the series $\Sigma a_n x^n$ converges at $x = c$, then it is absolutely convergent for $|x| < |c|$; if the series diverges at $x = c$, then it is divergent for $|x| > |c|$.

Proof. If $\Sigma a_n c^n$ is convergent, then $|a_n c^n| < K$, and

$$|a_n x^n| = |a_n c^n| \left|\frac{x}{c}\right|^n < K \left|\frac{x}{c}\right|^n.$$

Hence $\Sigma a_n x^n$ is absolutely convergent if $|x| < |c|$ by comparison. On the other hand, if $\Sigma a_n c^n$ is divergent and $|x_1| > |c|$, then $\Sigma a_n x_1^n$ must diverge; for if it converges then the series must converge when $x = c$, since $|c| \leqslant |x_1|$.

A direct consequence of the above theorem is as follows.

Theorem If $\Sigma a_n x^n$ is a power series which is not convergent everywhere or only at $x = 0$, then there is a positive number ρ such that $\Sigma a_n x^n$ is absolutely convergent for $|x| < \rho$ and divergent for $|x| > \rho$.

The number ρ is called the *radius of convergence*. When x is real the interval $(-\rho, \rho)$ is called the *interval of convergence*; when x is complex $|x| = \rho$ is the *circle of convergence*.

Suppose the power series converges at $x = c$; then it is absolutely convergent for $|x| < |c| = r$. Since we have assumed that the series does not converge everywhere, r must be bounded above and hence $\rho = \sup r$.

We frequently use the ratio test to find the radius of convergence of a given power series. Then $\rho = \lim_{n \to \infty} |a_n/a_{n+1}|$ when this limit exists.

Example 1 $\Sigma x^n/n^3$ is absolutely convergent for $|x| \leqslant 1$ and divergent for $|x| > 1$; that is $\rho = 1$ (See Example 2, Section 7.2 and Example 1, Section 7.4).

Example 2 $\Sigma x^{4n}/n$; here again $\rho = 1$. We saw in Example 2, Section 7.4 that this series is absolutely convergent for $|x| < 1$, and converges at each point of the circle of convergence except $x = \pm 1$ and $\pm i$.

Example 3 $\Sigma x^n/n!$ is absolutely convergent for all x, that is $\rho = \infty$. (See Example 3, Section 7.2)

Example 4 $\Sigma n! x^n$ converges for $x = 0$ only, that is $\rho = 0$.

For, using the ratio test we have

$$\frac{|n!x^n|}{|(n+1)!x^{n+1}|} = \frac{1}{n+1} \cdot \frac{1}{|x|} \to 0 \qquad \text{as } n \to \infty.$$

Example 5 $\Sigma a_n x^n$, $\Sigma n a_n x^{n-1}$ and $\Sigma [a_n/(n+1)] x^{n+1}$ all have the same radius of convergence.

Let the radii of convergence of the first two series be ρ and ρ_1, and suppose ρ is finite and non-zero. Since

$$|a_n x^n| \leqslant |n a_n x^n| = |x|\,|n a_n x^{n-1}| \qquad \text{for } n \geqslant 1,$$

$\Sigma a_n x^n$ is absolutely convergent when $\Sigma n a_n x^{n-1}$ is absolutely convergent by the comparison test. Therefore

$$\rho \geqslant \rho_1. \tag{1}$$

Now let $0<\alpha<\rho$, then $\Sigma a_n\alpha^n$ is convergent and $|a_n\alpha^n|$ is bounded, say $|a_n\alpha^n|\leqslant A$. Further

$$|na_nx^{n-1}| = |na_n\alpha^{n-1}|\left|\frac{x}{\alpha}\right|^{n-1} \leqslant An\left|\frac{x}{\alpha}\right|^{n-1},$$

and Σna_nx^{n-1} is absolutely convergent by comparison with the convergent series $\Sigma n(x/\alpha)^{n-1}$ for $|x|<\alpha$.

Hence $\rho_1\geqslant\alpha$, and since this is true for any α in $0<\alpha<\rho$, we have

$$\rho_1\geqslant\rho. \tag{2}$$

Combining inequalities (1) and (2) shows that $\rho_1=\rho$. If $\rho=\infty$, inequality (2) shows $\rho_1=\infty$. Since a_nx^n is the derivative of $a_nx^{n+1}/(n+1)$, the above result can be applied to the series

$$\Sigma\ \frac{a_n}{n+1}\,x^{n+1} \quad \text{and} \quad \Sigma\ a_nx^n.$$

7.8 Continuity of the sum function

Theorem If the power series

$$s(x) = \sum_{n=0}^{\infty} a_nx^n$$

has the radius of convergence ρ, then the sum function $s(x)$ is a continuous function of x for $|x|<\rho$.

Proof. Since the power series converges for $|x|<\rho$,

$$\forall\epsilon>0;\exists N.\ |s(x)-s_n(x)|<\tfrac{1}{3}\epsilon \qquad \text{for } n\geqslant N,$$

for each fixed x in $|x|<\rho$.

Let x_0 be any point of $|x|<\rho$; then in particular

$$|s(x_0)-s_N(x_0)|<\tfrac{1}{3}\epsilon.$$

Since $s_N(x)=\sum_{n=0}^{N} a_nx^n$ is a polynomial in x of degree N, it is a continuous function of x in the neighbourhood of $x=x_0$. That is, there is a $\delta>0$ such that

$$|s_N(x)-s_N(x_0)|<\tfrac{1}{3}\epsilon \qquad \text{for } |x-x_0|<\delta.$$

Hence, using these inequalities we have

$$\begin{aligned}|s(x)-s(x_0)| &\leqslant |s(x)-s_N(x)|+|s_N(x)-s_N(x_0)|+|s(x_0)-s_N(x_0)|\\ &<\epsilon \qquad \text{for } |x-x_0|<\delta.\end{aligned}$$

Thus $s(x)$ is continuous at x_0, and since x_0 is any point of $|x| < \rho$ we have shown that $s(x)$ is continuous in x for $|x| < \rho$.

Abel's theorem If $s(x) = \sum_{n=0}^{\infty} a_n x^n$ is convergent for $|x| < 1$ and Σa_n is convergent to a sum A, then $\lim_{x \to 1-0} s(x) = A$.

Proof. Since $A_n = \sum_{r=0}^{n} a_r$ converges to A, the partial sums A_n are bounded, that is $|A_n| < M$ for all n; also, given $\epsilon > 0$, $|A - A_n| < \frac{1}{2}\epsilon$ for $n \geqslant N$. In Exercises 7(c), No. 3, we saw that

$$(1-x)^{-1} \sum_{n=0}^{\infty} a_n x^n = \sum_{n=0}^{\infty} A_n x^n;$$

hence

$$\begin{aligned}(1-x)^{-1}\{s(x) - A\} &= (1-x)^{-1} \sum_{n=0}^{\infty} a_n x^n - A \sum_{n=0}^{\infty} x^n \\ &= \sum_{n=0}^{\infty} (A_n - A)x^n.\end{aligned}$$

Then

$$\begin{aligned}|s(x) - A| &\leqslant |1-x| \left\{ \sum_{n=0}^{N-1} |A - A_n| x^n + \sum_{n=N}^{\infty} |A - A_n| x^n \right\}, \\ &< 2M(1 - x^N) + \tfrac{1}{2}\epsilon x^N \qquad (0 < x < 1) \\ &< \tfrac{1}{2}\epsilon + \tfrac{1}{2}\epsilon = \epsilon \qquad \text{for } 0 < 1 - x < \delta = \frac{\epsilon}{4M}.\end{aligned}$$

Therefore $s(1-0) = \lim_{x \to 1-0} s(x) = A$.

7.9 Integration and differentiation of power series

Theorem If $s(x) = \sum_{n=0}^{\infty} a_n x^n$ is absolutely convergent for $|x| < \rho$, then

(a) $\displaystyle\int_a^b s(x)\,dx = \sum_{n=0}^{\infty} \int_a^b a_n x^n\,dx$ for $-\rho < a < x < b < \rho$;

(b) $\displaystyle s'(x) = \sum_{n=1}^{\infty} n a_n x^{n-1}$ for $|x| < \rho$.

In other words, provided we remain within the interval of convergence, a power series may be integrated or differentiated term by term and the resulting series has the expected sum.

Proof. We have seen (Example 5, Section 7.7) that the series $\Sigma a_n x^n$, $\Sigma n a_n x^{n-1}$ and $\Sigma a_n x^{n+1}/(n+1)$ all have the same radius of convergence.

Let $\alpha = \max(|a|, |b|)$; then $\Sigma a_n \alpha^n$ converges to A (say).

$$\forall \epsilon > 0; \exists N. \left| A - \sum_{r=0}^{n} a_r \alpha^r \right| \leqslant \sum_{r=n+1}^{\infty} |a_r| \alpha^r < \epsilon \qquad \text{for } n \geqslant N.$$

Also

$$|s(x) - s_n(x)| \leqslant \sum_{r=n+1}^{\infty} |a_r x^r| \leqslant \sum_{r=n+1}^{\infty} |a_r| \alpha^r \qquad \text{for } |x| \leqslant \alpha.$$

Then

$$\left| \int_a^b s(x)\,\mathrm{d}x - \int_a^b s_n(x)\,\mathrm{d}x \right| \leqslant \int_a^b |s(x) - s_n(x)|\,\mathrm{d}x$$

$$< (b-a)\epsilon \qquad \text{for } n \geqslant N,$$

which proves (a).

Let $\sum_{n=0}^{\infty} n a_n x^{n-1} = f(x)$; then $f(x)$ is a continuous function of x for $|x| < \rho$ and using (a),

$$\int_a^x f(t)\,\mathrm{d}t = \sum_{n=0}^{\infty} (a_n x^n - a_n a^n) \qquad (\text{where } -\rho < a < x < \rho)$$

$$= s(x) - s(a).$$

Differentiating, we have $f(x) = s'(x)$.

See also Exercises 7(d), No. 3.

Example 1 The geometric series $(1+x)^{-1} = \sum_{n=0}^{\infty} (-1)^n x^n$ is absolutely convergent for $|x| < 1$. Integration term by term gives

$$\log(1+x) = \sum_{n=0}^{\infty} \frac{(-1)^n x^{n+1}}{n+1} = \sum_{n=1}^{\infty} \frac{(-1)^{n-1} x^n}{n} \qquad \text{for } |x| < 1.$$

Abel's theorem (Section 7.8) shows that this result holds when $x = 1$.

Example 2 If the series $(1-x)^{-1} = \sum_{n=0}^{\infty} x^n$ is differentiated term by term we obtain

$$(1-x)^{-2} = \sum_{n=0}^{\infty} n x^{n-1} = \sum_{n=0}^{\infty} (n+1) x^n, \qquad |x| < 1.$$

Differentiating this series gives

$$(1-x)^{-3} = \sum_{n=0}^{\infty} \tfrac{1}{2}(n+1)(n+2) x^n \qquad \text{for } |x| < 1.$$

Example 3 Starting with

$$(1-x^2)^{-1} = \sum_{n=0}^{\infty} x^{2n}, \qquad |x| < 1,$$

we obtain by integration

$$\text{th}^{-1}\, x = \sum_{n=0}^{\infty} \frac{x^{2n+1}}{2n+1}, \qquad |x| < 1.$$

Example 4 If $A(x) = \sum_{n=0}^{\infty} a_n x^n$ and $B(x) = \sum_{n=0}^{\infty} b_n x^n$ are both convergent for $|x| < \rho$ and $A(x) = B(x)$ for all x in $|x| < \rho$, then $a_n = b_n$ for all n.

Since $A(x) = B(x)$ in $|x| < \rho$, we have, putting $x = 0$,

$$a_0 = b_0.$$

Differentiating with respect to x and putting $x = 0$ gives

$$\sum_{n=0}^{\infty} n a_n x^{n-1} = \sum_{n=0}^{\infty} b_n x^{n-1} \quad \text{and} \quad a_1 = b_1.$$

Further differentiation gives, for $|x| < \rho$,

$$A^{(k)}(x) = B^{(k)}(x), \qquad k = 1, 2, 3, \ldots .$$

Putting $x = 0$, we have $a_n = b_n$, $n = 0, 1, 2, 3, \ldots$.

Note. This implies that if $\sum_{n=0}^{\infty} a_n x^n = 0$ for all x in $|x| < \rho$, then $a_n = 0$ for all n.

7.10 Multiplication of power series

Theorem If

$$A(x) = \sum_{n=0}^{\infty} a_n x^n, \qquad |x| < \rho_1$$

and

$$B(x) = \sum_{n=0}^{\infty} b_n x^n, \qquad |x| < \rho_2,$$

then

$$A(x)B(x) = \sum_{n=0}^{\infty} c_n x^n \qquad \text{for } |x| < \rho = \min(\rho_1, \rho_2),$$

where $c_n = \sum_{r=0}^{n} a_r b_{n-r}$.

Proof. Since the power series for $A(x)$ and $B(x)$ are both absolutely convergent for $|x| < \rho$, the result follows from the theorem of Section 7.6.

Example 1 In Example 3 of the previous section, we found

$$\text{th}^{-1}\,x = \sum_{n=0}^{\infty} \frac{x^{2n+1}}{2n+1}, \qquad |x|<1.$$

Then

$$(1-x)^{-1}\,\text{th}^{-1}\,x = (1+x+x^2+\cdots)(x+\tfrac{1}{3}x^3+\tfrac{1}{5}x^5+\cdots)$$
$$= x+x^2+(1+\tfrac{1}{3})x^3+(1+\tfrac{1}{3})x^4+(1+\tfrac{1}{3}+\tfrac{1}{5})x^5$$
$$+(1+\tfrac{1}{3}+\tfrac{1}{5})x^6+\cdots, \qquad |x|<1.$$

Example 2 If the series $\sum_{n=0}^{\infty} a_n$, $\sum_{n=0}^{\infty} b_n$, $\sum_{n=0}^{\infty} c_n$, where

$$c_n = \sum_{r=0}^{n} a_r b_{n-r},$$

are convergent to sums A, B and C respectively, then $C = AB$.

Consider the power series $A(x) = \sum_{n=0}^{\infty} a_n x^n$, $B(x) = \sum_{n=0}^{\infty} b_n x^n$ and $C(x) = \sum_{n=0}^{\infty} c_n x^n$; all three series are absolutely convergent for $|x|<1$.

Therefore $\qquad A(x)B(x) = C(x) \qquad$ for $|x|<1$.

Also the three series converge at $x = 1$, so Abel's theorem may be used to give the required result.

Example 3 If $A(x) = \sum_{n=0}^{\infty} a_n x^n$ is absolutely convergent for $|x|<\rho$, and $a_0 \neq 0$, then there is a $\delta > 0$ such that

$$\frac{1}{A(x)} = \sum_{n=0}^{\infty} d_n x^n, \qquad \text{for } |x|<\delta,$$

where the coefficients are given by

$$d_0 = 1/a_0,\, d_1 = -a_1/a_0^2,\, d_2 = (a_1^2 - a_0 a_2)/a_0^3, \ldots .$$

The coefficients are found successively by equating to zero coefficients of powers of x in the product of the two series. (See Example 4, Section 7.9.)

$$1 = (a_0 + a_1 x + a_2 x^2 + a_3 x^3 + \cdots)(d_0 + d_1 x + d_2 x^2 + d_3 x^3 + \cdots)$$
$$= a_0 d_0 + (a_0 d_1 + a_1 d_0)x + (a_0 d_2 + a_1 d_1 + a_2 d_0)x^2 + \cdots .$$

Hence

$$a_0 d_0 = 1,$$
$$a_0 d_1 + a_1 d_0 = 0,$$
$$a_0 d_2 + a_1 d_1 + a_2 d_0 = 0,$$

and so on.

Example 4 Let

$$S(x) = x - \frac{x^3}{3!} + \frac{x^5}{5!} - \cdots \quad \text{and} \quad C(x) = 1 - \frac{x^2}{2!} + \frac{x^4}{4!} - \cdots,$$

each series being absolutely convergent for all real x. Then

$$\frac{S(x)}{C(x)} = \sum_{n=0}^{\infty} d_n x^n, \qquad |x| < \rho$$

for some $\rho > 0$, where

$$x - \frac{x^3}{3!} + \frac{x^5}{5!} - \cdots = \left(1 - \frac{x^2}{2!} + \frac{x^4}{4!} - \cdots\right)(d_0 + d_1 x + d_2 x^2 + \cdots).$$

Multiplying out the series on the right of this equation and equating coefficients of like powers of x gives

$$d_0 = 0, \quad d_1 = 1, \quad d_2 = d_4 = d_6 = 0$$

and

$$d_3 = \tfrac{1}{3}, \quad d_5 = \tfrac{2}{15}, \quad d_7 = \tfrac{17}{315}.$$

That is

$$\frac{S(x)}{C(x)} = x + \tfrac{1}{3}x^3 + \tfrac{2}{15}x^5 + \tfrac{17}{315}x^7 + \cdots.$$

7.11 Substitution of one power series into another

Theorem Let $f(x) = \sum_{n=0}^{\infty} a_n x^n$ and $g(x) = \sum_{n=0}^{\infty} b_n x^n$ have positive radii of convergence, ρ_1 and ρ_2 respectively; then if $|b_0| < \rho_1$, there is a $\delta > 0$ such that for $|x| < \delta$, $f(g(x))$ is given by the formal substitution of $\Sigma b_n x^n$ into $\Sigma a_n x^n$.

That is

$$f(g(x)) = a_0 + a_1(b_0 + b_1 x + b_2 x^2 + \cdots) + a_2(b_0 + b_1 x + b_2 x^2 + \cdots)^2 + \cdots;$$

and $f(g(0)) = f(b_0) = \Sigma a_n (b_0)^n$ converges since $|b_0| < \rho_1$.

Example Let

$$f(x) = \sum_{n=0}^{\infty} \frac{x^n}{n!} \quad \text{and} \quad g(x) = \sum_{n=0}^{\infty} \frac{(-1)^n x^{2n+1}}{(2n+1)!};$$

both series are absolutely convergent for all real x. Then

$$f(g(x)) = 1+\left(x-\frac{x^3}{3!}+\frac{x^5}{5!}-\cdots\right)+\frac{1}{2!}\left(x-\frac{x^3}{3!}+\frac{x^5}{5!}-\cdots\right)^2$$
$$+\frac{1}{3!}\left(x-\frac{x^3}{3!}+\cdots\right)^3$$
$$+\frac{1}{4!}\left(x-\frac{x^3}{3!}+\cdots\right)^4+\frac{1}{5!}\left(x-\frac{x^3}{3!}+\cdots\right)^5+\cdots$$
$$= 1+x+\tfrac{1}{2}x^2+\left(\frac{1}{4!}-\frac{2}{2!3!}\right)x^4+\left(\frac{1}{5!}-\frac{3}{3!3!}+\frac{1}{5!}\right)x^5+\cdots$$
$$= 1+x+\tfrac{1}{2}x^2-\tfrac{1}{8}x^4-\tfrac{1}{15}x^5+\cdots;$$

EXERCISES 7(d)

1. Test the following series for convergence when x is real

(a) $\sum\frac{(-1)^n x^n}{(2n+1)3^n}$ (b) $\sum n\left(\frac{x}{5}\right)^{2n}$ (c) $\sum\frac{n^3x^n}{n!}$

(d) $\sum n^2(x-1)^n$ (e) $\sum\frac{n+1}{x^n}$ (f) $\sum\frac{(-1)^n(x-4)^n}{n^3}$

(g) $\sum\frac{1.3.5\cdots(2n-1)}{2.4.6\cdots(2n)}x^n$ (h) $\sum\frac{1.3.5\cdots(2n-1)}{2.4.6\cdots(2n)}\frac{x^n}{n}$

(i) $\sum\frac{(n!)^2x^n}{(2n)!}$ (j) $\sum\frac{3.7.11\cdots(4n-1)}{4.8.12\cdots(4n)}x^n$.

2. Discuss the convergence of the following power series for complex z:

(a) $\sum\frac{z^n}{n}$ (b) $\sum nz^{2n}$ (c) $\sum\frac{z^n}{n^2}$

(d) $\sum\frac{z^{4n}}{n\sqrt{n}}$ (e) $\sum\frac{(-1)^n z^{2n}}{n}$

3. Prove that, if $|x|<\rho$ and $|a|<\rho$, then

$$\left|\frac{x^n-a^n}{x-a}-na^{n-1}\right|\leqslant\tfrac{1}{2}|x-a|n(n-1)\rho^{n-2},\qquad n=2,3,4,\ldots.$$

If $f(x)=\sum_{n=0}^{\infty}a_nx^n$ for $|x|<\rho$ and a is any number such that $|a|<\rho$ show, by considering

$$\left|\frac{f(x)-f(a)}{x-a}-\sum_{n=1}^{\infty}na_na^{n-1}\right|,$$

that f has a derivative at a equal to Σna_na^{n-1}.

4. Starting with

$$(1+x^2)^{-1} = \sum_{n=0}^{\infty} (-1)^n x^{2n}, \qquad |x|<1,$$

show that

$$\tan^{-1} x = \sum_{n=0}^{\infty} \frac{(-1)^n x^{2n+1}}{2n+1}, \qquad |x|<1.$$

show also that

$$\int_0^1 \frac{\tan^{-1} x}{x}\,dx = 1 - \frac{1}{3^2} + \frac{1}{5^2} - \cdots .$$

5. Assuming the binomial expansion

$$(1-x^2)^{-1/2} = \sum_{n=0}^{\infty} \frac{(2n)!}{(n!)^2} (\tfrac{1}{2}x)^{2n},$$

for $|x|<1$, show that

$$\sin^{-1} x = \sum_{n=0}^{\infty} \frac{(2n)!}{(n!)^2 2^{2n}} \frac{x^{2n+1}}{2n+1}, \qquad |x| \leqslant 1.$$

6. Show that if $f(x) = \log\,[\sqrt{(1+x)} + \sqrt{(1-x)}]$, then

$$f'(x) = \frac{1}{4x}\left(\frac{1}{\sqrt{(1-x^2)}} - 1\right).$$

Show that

$$\log\,\{\sqrt{(1+x)} + \sqrt{(1-x)}\}$$

$$= \log 2 - \sum_{n=1}^{\infty} \frac{1.3.5\cdots(2n-1)}{2.4.6\cdots(2n)} \frac{x^{2n}}{4n}, \qquad |x| \leqslant 1.$$

7. Assuming that

$$\tan x = x + \tfrac{1}{3}x^3 + \tfrac{2}{15}x^5 + \tfrac{17}{315}x^7 + \cdots,$$

find the series for $\sec^2 x$ and $\log \sec x$ as far as the term in x^6.

8. Show that if

$$C(x) = 1 - \frac{x^2}{2!} + \frac{x^4}{4!} - \cdots,$$

then

$$\frac{1}{C(x)} = 1 + \tfrac{1}{2}x^2 + \tfrac{5}{24}x^4 + \cdots$$

for sufficiently small values of x.

9. Show that, if $|z| < 1$,

$$\left(\sum_{n=1}^{\infty} z^n \sin n\theta\right)\left(\sum_{n=0}^{\infty} z^n \cos n\theta\right) = \frac{1}{2}\sum_{n=0}^{\infty}(n+1)z^n \sin n\theta.$$

10. Show that if

$$f(x) = x - \tfrac{1}{2}x^2 + \tfrac{1}{3}x^3 - \tfrac{1}{4}x^4 + \cdots$$

and

$$g(x) = x - \frac{x^3}{3!} + \frac{x^5}{5!} - \cdots,$$

then $f(g(x)) = x - x^2/2 + x^3/6 - x^4/12 + x^5/24 + \cdots$.

11. Prove that if Σa_n^2, Σb_n^2 are convergent and a_n, b_n are positive numbers, then $\Sigma a_n b_n$ is convergent and

$$\Sigma a_n b_n \leqslant (\Sigma a_n^2)^{1/2}(\Sigma b_n^2)^{1/2}.$$

CHAPTER 8

Taylor's Theorem and Series

8.1 Taylor's theorem

In the previous chapter we saw that a power series may be differentiated term by term and that the derived series has the same radius of convergence as the original series. Thus, if $f(x) = \Sigma_{n=0}^{\infty} a_n x^n$ is convergent for $|x| < \rho$, we have by repeated differentiation

$$f'(x) = \sum_{n=1}^{\infty} n a_n x^{n-1},$$
$$f''(x) = \sum_{n=2}^{\infty} n(n-1) a_n x^{n-2},$$

and generally

$$f^{(k)}(x) = \sum_{n=k}^{\infty} n(n-1)(n-2)\cdots(n-k+1) a_n x^{n-k},$$
$$k = 1, 2, 3, \ldots.$$

Hence, putting $x = 0$ gives $a_0 = f(0)$,

$$a_1 = f'(0),$$
$$a_2 = f''(0)/2!,$$

and generally $\quad a_k = f^{(k)}(0)/k!, \qquad k = 1, 2, 3, \ldots.$

Thus we can write

$$f(x) = \sum_{n=0}^{\infty} \frac{f^{(n)}(0)}{n!} x^n.$$

This shows that if we start with the power series $\Sigma a_n x^n$, then the coefficients must have the values just found.

We now consider this problem from another point of view – under what conditions can a given function $f(x)$ be expanded in a power series?

Taylor's theorem If $f(x)$ and $f^{(r)}(x), r = 1, 2, \ldots, n-1$ are continuous in an interval $[a, b]$ and $f^{(n)}(x)$ exists in (a, b), then

$$\begin{aligned} f(b) &= f(a) + (b-a)f'(a) + \frac{(b-a)^2}{2!} f''(a) + \cdots \\ &\quad + \frac{(b-a)^{n-1}}{(n-1)!} f^{(n-1)}(a) + R_n \\ &= \sum_{r=0}^{n-1} \frac{f^{(r)}(a)}{r!} (b-a)^r + R_n, \end{aligned}$$

where

$$R_n = \frac{(b-a)^n}{n!} f^{(n)}(\xi), \qquad a < \xi < b.$$

Proof. We prove this theorem by applying Rolle's theorem (Section 4.7) to the auxiliary function

$$G(x) = F_n(x) - \left(\frac{b-x}{b-a}\right)^n F_n(a),$$

where

$$\begin{aligned} F_n(x) &= f(b) - f(x) - (b-x)f'(x) - \frac{(b-x)^2}{2!} f''(x) - \cdots \\ &\quad - \frac{(b-x)^{n-1}}{(n-1)!} f^{(n-1)}(x). \end{aligned}$$

$G(a) = G(b) = 0$; since $f^{(r)}(x)$ $(r = 0, 1, 2, \ldots, n-1)$ are continuous in $[a, b]$, $F_n(x)$ is continuous and consequently $G(x)$ is also continuous in $[a, b]$. Since $f^{(n)}(x)$ exists in (a, b), so also does $G'(x)$ exist in (a, b). Therefore G satisfies the conditions of Rolle's theorem. That is there exists a point ξ, $a < \xi < b$, such that $G'(\xi) = 0$. Now

$$G'(x) = \frac{n(b-x)^{n-1}}{(b-a)^n} F_n(a) - \frac{(b-x)^{n-1}}{(n-1)!} f^{(n)}(x).$$

Hence

$$F_n(a) = R_n = \frac{(b-a)^n}{n!} f^{(n)}(\xi), \qquad a < \xi < b.$$

This remainder is known as the *Lagrange* form. It is sometimes more convenient to take the interval to be $[a, a+h]$ in place of $[a, b]$. In this case the result can be written

$$f(a+h) = \sum_{r=0}^{n-1} \frac{h^r}{r!} f^{(r)}(a) + \frac{h^n}{n!} f^{(n)}(a + \theta h), \qquad 0 < \theta < 1.$$

Note Taking $n = 1$ gives the mean value theorem (Section 4.8); Taylor's theorem is also known as the nth mean value theorem.

When the power series is centred on the origin, that is, when $a = 0$ and the expansion is of the form

$$f(x) = \sum_{r=0}^{n-1} \frac{x^r}{r!} f^{(r)}(0) + R_n,$$

the result is known as Maclaurin's theorem.

Another useful form of the remainder, due to *Cauchy*, may be obtained by applying the mean value theorem to the function $F_n(x)$ defined above. Then

$$F_n(b) - F_n(a) = (b-a)F'_n(\xi), \qquad a < \xi < b.$$

Hence $R_n = F_n(a) = (b-a)F'_n(\xi)$, (since $F_n(b) = 0$)

$$= \frac{(b-a)(b-\xi)^{n-1}}{(n-1)!} f^{(n)}(\xi), \qquad a < \xi < b.$$

Writing $b = a + h$ and $\xi = a + \theta h$, we obtain for the remainder

$$R_n = \frac{(1-\theta)^{n-1} h^n}{(n-1)!} f^{(n)}(a + \theta h), \qquad 0 < \theta < 1.$$

Example 1 If $f(x) = e^x$, then $f^{(n)}(x) = e^x$, $n = 1, 2, 3, \ldots$. Hence

$$e^b = e^a \sum_{r=0}^{n-1} \frac{(b-a)^r}{r!} + R_n,$$

where

$$R_n = \frac{(b-a)^n}{n!} e^{\xi}, \qquad a < \xi < b.$$

Example 2 $f(x) = \log x, \qquad 0 < a \leqslant x \leqslant a + h.$

$f'(x) = 1/x$ and generally

$$f^{(r)}(x) = \frac{(-1)^{r-1}(r-1)!}{x^r}, \qquad r = 1, 2, 3, \ldots.$$

Hence

$$\log(a+h) = \log a + \sum_{r=1}^{n-1} \frac{h^r}{r!} \frac{(-1)^{r-1}(r-1)!}{a^r} + R_n$$

$$= \log a + \sum_{r=1}^{n-1} (-1)^{r-1} \left(\frac{h}{a}\right)^r + R_n,$$

where

$$R_n = (-1)^{n-1} \frac{(1-\theta)^{n-1} h^n}{(a+h)^n}, \qquad 0 < \theta < 1.$$

8.2 Young's form of Taylor's theorem

Theorem If $f(x)$ and $f^{(r)}(x), r = 1, 2, \ldots, n-1$ are continuous in an interval $[a, a+h]$ and $f^{(n)}(a)$ exists, then

$$f(a+h) = \sum_{r=0}^{n-1} \frac{h^r}{r!} f^{(r)}(a) + \frac{h^n}{n!}(f^{(n)}(a) + \epsilon),$$

where $\epsilon \to 0$ as $h \to 0$.

Proof. Let

$$F(h) = f(a+h) - \sum_{r=0}^{n} \frac{h^r}{r!} f^{(r)}(a) \quad \text{and} \quad G(h) = \frac{h^n}{n!};$$

then $\epsilon = F(h)/G(h)$ and

$$\lim_{h \to 0} \epsilon = \lim_{h \to 0} \frac{F(h)}{G(h)}$$

$$= \lim_{h \to 0} \frac{F^{(n)}(h)}{G^{(n)}(h)}, \qquad \text{using Theorem 3 of Section 4.10.}$$

Now

$$F^{(n)}(h) = f^{(n)}(a+h) - f^{(n)}(a) \quad \text{and} \quad G^{(n)}(h) = 1,$$

and so

$$\lim_{h \to 0} \epsilon = 0.$$

Using this form of Taylor's theorem we can establish the general conditions for a function f to have an extreme value (Section 4.6).

Theorem If $f^{(r)}(a) = 0$ for $r = 0, 1, 2, \ldots, n-1$ and $f^{(n)}(a) \neq 0$, then if n is even, f has an extreme value at $x = a$ which is a minimum if $f^{(n)}(a) > 0$ and a maximum if $f^{(n)}(a) < 0$; if n is odd f has no extreme value at $x = a$.

Proof. With the given conditions we can write

$$f(a+h) - f(a) = \frac{h^n}{n!}(f^{(n)}(a) + \epsilon), \qquad \text{where } \epsilon \to 0 \text{ as } h \to 0.$$

Since $f^{(n)}(a) \neq 0$, we can find an h sufficiently small, $|h| < \delta$, for $f^{(n)}(a) + \epsilon$ to have the sign of $f^{(n)}(a)$ in $|x - a| < \delta$. Then the sign of $f(a+h) - f(a)$ changes with the sign of h if n is odd. Hence no extreme value if n is odd.

If n is even, $h^n > 0$ for all h in the neighbourhood of a, so $f(a+h) - f(a)$ preserves the sign of $f^{(n)}(a)$ throughout $|x - a| < \delta$, and the result follows.

EXERCISES 8(a)

1. By applying the mean value theorem to the function $F_n(x)$ defined in Section 8.1, obtain Taylor's theorem with the Cauchy form of remainder.

2. Prove Taylor's theorem with the Lagrange form of remainder by applying Cauchy's mean value theorem to the functions $F_n(x)$ (defined in Section 8.1) and $G(x) = (b - x)^n$.

3. Given that f is a polynomial of degree k and R_n is the remainder after n terms of the Taylor expansion of f, show that $R_n = 0$ for $n \geqslant k + 1$.

4. Given that $f^{(n+1)}(x)$ is continuous at $x = a$, and

$$R_n = \frac{h^n}{n!} f^{(n)}(a + \theta h), \qquad 0 < \theta < 1,$$

is the remainder after n terms of the Taylor expansion of f, show that $\theta \to 1/(n + 1)$ as $h \to 0$.

5. By applying Rolle's theorem to the auxiliary function

$$H(x) = f(a + h) - f(x) - (a + h - x)f'(x) - \cdots - \frac{(a + h - x)^{n-1}}{(n - 1)!} f^{(n-1)}(x) - (a + h - x)^p K,$$

where p is a positive integer and K, independent of x, is such that $H(a) = 0$, obtain the Schlömilch–Roche form of remainder,

$$\frac{x^n(1 - \theta)^{n-p}}{(n - 1)!p} f^{(n)}(\theta x), \qquad 0 < \theta < 1.$$

What conditions must one impose on f?

6. Show that

$$\frac{1}{x + a} = \frac{1}{x} - \frac{a}{x^2} + \frac{a^2}{x^3} - \frac{a^3}{x^4} + \cdots, \qquad |x| < |a|.$$

7. The remainder after n terms of the Taylor expansion of a function f is

$$\frac{h^n}{n!} f^{(n)}(a + \theta_n h), \qquad 0 < \theta_n < 1.$$

If f is a polynomial of degree $n + 2$ and $f^{(n+1)}$ does not vanish show that

$$\theta_n \approx \frac{1}{n+1} + \frac{nh}{2(n+1)^2(n+2)} \frac{f^{(n+2)}(a)}{f^{(n+1)}(a)},$$

if $|h|$ is small.

8.3 Taylor and Maclaurin series

If f has derivatives of all orders in an interval $|x-a|<\delta$, and if $|h|<\delta$, then

$$f(a+h) = \sum_{r=0}^{n-1} \frac{h^r}{r!} f^{(r)}(a) + \frac{h^n}{n!} f^{(n)}(a+\theta h), \qquad 0<\theta<1,$$

$$= S_n + R_n, \qquad \text{for any integer } n.$$

Then, if $R_n \to 0$ as $n \to \infty$ when $|h|<\rho<\delta$, the resulting infinite series must converge to the sum function $f(a+h)$, and is called the *Taylor series* of the function. That is

$$f(a+h) = \sum_{n=0}^{\infty} \frac{h^n}{n!} f^{(n)}(a) \qquad \text{for } |h|<\rho.$$

If we write $h = x-a$, the Taylor series for f is

$$f(x) = \sum_{n=0}^{\infty} \frac{(x-a)^n}{n!} f^{(n)}(a) \qquad \text{for } |x-a|<\rho.$$

When $a=0$,

$$f(x) = \sum_{n=0}^{\infty} \frac{x^n}{n!} f^{(n)}(0) \qquad \text{for } |x|<\rho,$$

and this is known as the *Maclaurin series* for f.

It is important to notice that the convergence of the infinite series does not imply that $R_n \to 0$ as $n \to \infty$ (See Exercises 8(b) No. 10); but if $R_n \to 0$ as $n \to \infty$, it necessarily follows that the series converges. Thus, for the series to represent the function f, that is, for the sum function of the series to be f, then R_n must tend to zero as $n \to \infty$. On the other hand there is little point in considering the behaviour of R_n for values of x for which the series diverges. Consequently we must first determine the range of values of x for which the series converges, and then investigate the behaviour of R_n for these values of x.

In the following sections we find the Maclaurin series and some properties of the elementary functions of analysis. For the definitions of the logarithmic and exponential functions see, for example, Hardy (1952), Chapter 10.

8.4 The circular functions

In Example 3, Section 4.4 we found that if $f(x) = \sin x$, then $f^{(n)}(x) = \sin(x + \frac{1}{2}n\pi)$, $n = 1, 2, 3, \ldots$. Hence, for $k = 1, 2, 3, \ldots$,

$$f^{(2k)}(0) = 0 \quad \text{and} \quad f^{(2k-1)}(0) = (-1)^{k-1}.$$

Consequently

$$\sin x = x - \frac{x^3}{3!} + \frac{x^5}{5!} - \cdots + (-1)^{n-2}\frac{x^{2n-3}}{(2n-3)!} + R_{2n-1},$$

where

$$R_{2n-1} = \frac{x^{2n-1}}{(2n-1)!}\sin\left[\theta x + \tfrac{1}{2}(2n-1)\pi\right], \qquad 0 < \theta < 1.$$

Then

$$|R_{2n-1}| \leqslant \frac{|x|^{2n-1}}{(2n-1)!} \to 0 \qquad \text{as } n \to \infty$$

for any finite x, since the series $\Sigma\, x^{2n-1}/(2n-1)!$ is absolutely convergent for all x. Hence the Maclaurin series for $\sin x$ is given by

$$\sin x = \sum_{n=0}^{\infty} \frac{(-1)^n x^{2n+1}}{(2n+1)!} \qquad \text{for all } x.$$

Similarly

$$\cos x = \sum_{n=0}^{\infty} \frac{(-1)^n x^{2n}}{(2n)!} \qquad \text{for all } x.$$

8.5 The exponential and hyperbolic functions

Since the derivative of e^x is itself, we have

$$e^x = \sum_{r=0}^{n-1} \frac{x^r}{r!} + \frac{x^n}{n!}e^{\theta x}, \qquad 0 < \theta < 1.$$

For any x, $e^{\theta x} < e^x$ and $x^n/n! \to 0$ as $n \to \infty$, since it is the nth term of a convergent series (Example 3, Section 7.2). Therefore the Maclaurin series for the exponential function is

$$e^x = 1 + x + \frac{x^2}{2!} + \frac{x^3}{3!} + \cdots + \frac{x^n}{n!} + \cdots \text{ for all } x.$$

Theorem The exponential function increases more rapidly than any power of x; that is,

$$\lim_{x \to \infty} x^{\alpha} e^{-x} = 0 \text{ for all } \alpha.$$

Proof. Choose any $\beta > \alpha$, and let p be a positive integer such that $p - 1 < \beta \leqslant p$. Then
$$e^x > x^p/p! \qquad \text{for } x > 0,$$
$$> x^{\beta}/p! \qquad \text{for } x > 1.$$

Hence
$$x^{\alpha} e^{-x} < p!\, x^{\alpha-\beta}$$
$$\to 0 \text{ as } x \to \infty, \qquad \text{since } \beta > \alpha.$$

When x is real,

$$e^{ix} = \sum_{n=0}^{\infty} \frac{(-1)^n x^{2n}}{(2n)!} + i \sum_{n=0}^{\infty} \frac{(-1)^n x^{2n+1}}{(2n+1)!},$$
$$= \cos x + i \sin x.$$

Also
$$e^{-ix} = \cos x - i \sin x.$$

The circular functions can be defined for any real or complex x

$$\cos x = \frac{e^{ix} + e^{-ix}}{2} \quad \text{and} \quad \sin x = \frac{e^{ix} - e^{-ix}}{2i}.$$

In Section 2.2 the hyperbolic functions were defined in terms of the exponential function,

$$\operatorname{ch} x = \tfrac{1}{2}(e^x + e^{-x}) \quad \text{and} \quad \operatorname{sh} x = \tfrac{1}{2}(e^x - e^{-x}).$$

Using the series for e^x and e^{-x} we see that these functions are the even and odd parts respectively of the exponential function. Thus

$$\operatorname{ch} x = 1 + \frac{x^2}{2!} + \frac{x^4}{4!} + \cdots = \sum_{n=0}^{\infty} \frac{x^{2n}}{(2n)!},$$

$$\operatorname{sh} x = x + \frac{x^3}{3!} + \frac{x^5}{5!} + \cdots = \sum_{n=0}^{\infty} \frac{x^{2n-1}}{(2n+1)!},$$

both series being absolutely convergent for all x.

The hyperbolic tangent is defined to be

$$\operatorname{th} x = \frac{\operatorname{sh} x}{\operatorname{ch} x} = \frac{e^x - e^{-x}}{e^x + e^{-x}}$$
$$= \frac{1 - e^{-2x}}{1 + e^{-2x}} = \frac{e^{2x} - 1}{e^{2x} + 1}.$$

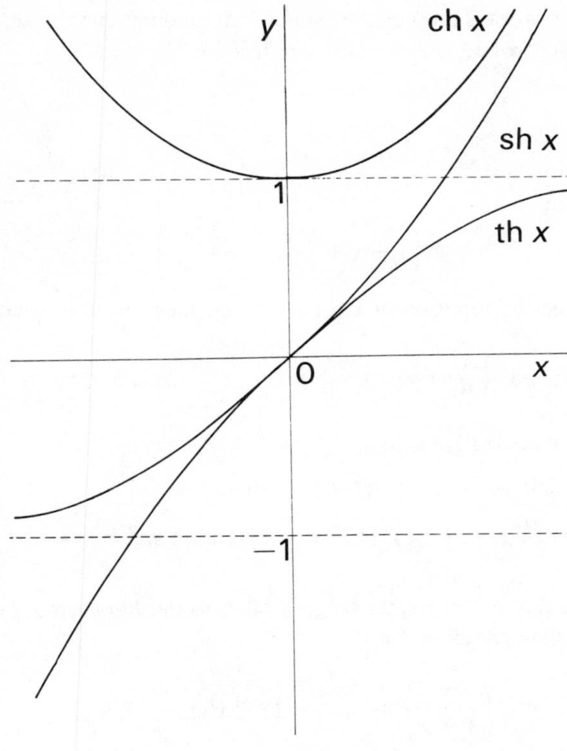

Fig. 5

It can be seen that th $x \to \pm 1$ as $x \to \infty$. The graphs of these functions are shown in Fig. 5.

The reciprocals of these hyperbolic functions are

$$\coth x = \frac{1}{\operatorname{th} x}, \quad \operatorname{sech} x = \frac{1}{\operatorname{ch} x}, \quad \operatorname{cosech} x = \frac{1}{\operatorname{sh} x}.$$

From the definitions we see that the circular and hyperbolic functions are related. Thus

$$\operatorname{ch} x = \cos ix \quad \text{and} \quad i \operatorname{sh} x = \sin ix,$$
$$\operatorname{ch} ix = \cos x \quad \text{and} \quad \operatorname{sh} ix = i \sin x,$$

Since $i^2 = -1$, this leads to a simple rule for obtaining purely algebraic hyperbolic identities from the corresponding trigonometric ones. The circular functions are replaced by the corresponding hyperbolic functions leaving the signs unchanged *except* the sign of a term which involves the product of two sines. Thus $\cos^2 x + \sin^2 x = 1$ becomes $\mathrm{ch}^2 x - \mathrm{sh}^2 x = 1$; $1 + \tan^2 x = \sec^2 x$ becomes $1 - \mathrm{th}^2 x = \mathrm{sech}^2 x$.

EXERCISES 8(b)

1. Show that

$$e^{-x}\left(1 + x + \frac{x^2}{2!} + \cdots + \frac{x^n}{n!}\right)$$

is a decreasing function of x for $x > 0$. Deduce that the equation

$$1 + x + \frac{x^2}{2!} + \cdots + \frac{x^n}{n!} = ae^x, \qquad 0 < a < 1,$$

has only one positive root.

2. Prove that if $y = \tan^{-1}[x/(1+x)]$, then

$$D^n y = (-1)^{n-1} 2^n (n-1)! \sin n\theta \sin^n \theta,$$

where $\cot \theta = 2x + 1$.

Prove that if the inverse tangent takes values between $-\frac{3}{4}\pi$ and $\frac{1}{4}\pi$, then for all real x,

$$\tan^{-1}\left(\frac{x}{1+x}\right) = \sum_{r=1}^{n-1} (-1)^{r-1} x^r \sin\frac{r\pi}{4} + R_n,$$

where $|R_n| < 2^n |x|^n / n$.

3. Show that

$$\sum_{r=n+1}^{\infty} \frac{1}{r!} < \frac{1}{n \,.\, n!}.$$

Deduce that e is irrational.

4. Prove that, if f' exists in $a < x < a + h$, then

$$f(a+h) = f(a) + (e^h - 1)e^{-\theta h} f'(a + \theta h), \qquad 0 < \theta < 1.$$

5. Show that

$$\sin(x+h) = \sin x + h\cos x - \frac{h^2}{2!}\sin x - \frac{h^3}{3!}\cos x + \cdots$$
$$+ (-1)^n \frac{h^{2n-1}}{(2n-1)!}\cos x + (-1)^n \frac{h^{2n}}{(2n)!}\sin(x + \theta h),$$

$0 < \theta < 1$.

6. Show that

$$\operatorname{ch} x = 1 + \frac{x^2}{2!} + \frac{x^4}{4!} + \cdots + \frac{x^{2n}}{(2n)!} + \frac{x^{2n+1}}{(2n+1)!} \operatorname{sh} \theta x, \qquad 0 < \theta < 1.$$

7. Show that

$$e^{ax} \cos bx = 1 + xr \cos \alpha + \frac{x^2 r^2}{2!} \cos 2\alpha + \cdots + \frac{x^n r^n}{n!} \cos n\alpha + \cdots,$$

where $r = \sqrt{(a^2 + b^2)}$ and $\alpha = \tan^{-1}(b/a)$.

8. Evaluate:

(a) $\lim_{\theta \to 0} \dfrac{1 - \cos 3\theta}{\cos \theta - \cos 2\theta}$ (b) $\lim_{x \to 0} \dfrac{e^x - \cos x - \sin x - x^2}{(2 + 2x + x^2) e^{-x} - 2}$

(c) $\lim_{x \to 0} \dfrac{2 - \sin^2 x - 2 \cos x}{x^4}$ (d) $\lim_{x \to 0} \dfrac{\sin x - \operatorname{sh} x + \frac{1}{3}x^3}{x(\operatorname{sh} x^2 + \cos x - \operatorname{ch} x)}$

9. Show that

$$\operatorname{sh} x \sin x = \frac{2x^2}{2!} - \frac{2^3 x^6}{6!} + \frac{2^5 x^{10}}{10!} - \cdots + \frac{(-1)^{n-1} 2^{2n-1} x^{4n-2}}{(4n-2)!} + \frac{(-1)^n 2^{2n+1} x^{4n+2}}{(4n+2)!} + M,$$

where $|M| < \operatorname{ch} x$.

10. Given $g(x) = \exp(-1/x^2), x \neq 0, g(0) = 0$, show that $g'(0) = 0$ and $x^3 g'(x) = 2g(x), x \neq 0$. Deduce that $g^{(n)}(0) = 0$, $n = 1, 2, 3, \ldots$. Show that the Maclaurin series for $f(x) = e^x + g(x)$ converges for all x but *not* to $f(x)$.

8.6 The logarithmic function

Since $\log x$ and its derivatives are all infinite at $x = 0$, this function does not possess a Maclaurin series. However, we can find a Taylor series of the form $\Sigma a_n (x - a)^n$ for any $a > 0$ (See Exercises 8(c) No. 3).

A more useful approach is to consider the function $\log(1 + x)$, which is continuous and possesses continuous derivatives at the origin. Let $f(x) = \log(1 + x)$; then

$$f^{(n)}(x) = \frac{(-1)^{n-1}(n-1)!}{(1+x)^n}, \qquad n = 1, 2, 3, \ldots.$$

Hence

$$\log(1+x) = x - \frac{x^2}{2} + \frac{x^3}{3} - \cdots + (-1)^n \frac{x^{n-1}}{n-1} + R_n,$$

where

$$R_n = (-1)^{n-1}\frac{x^n}{n}(1+\theta x)^{-n}, \qquad 0<\theta<1.$$

We have seen (Example 1, Section 7.2) that the series $\Sigma(-1)^{n-1}x^n/n$ converges for $-1<x\leqslant 1$, so we must show that $R_n \to 0$ as $n\to\infty$ for this range of values of x. Now, when $0<x\leqslant 1$,

$$|R_n| < \frac{x^n}{n} \to 0 \qquad \text{as } n\to\infty;$$

when $-1<x<0$ this inequality breaks down. Using Cauchy's form of the remainder,

$$R_n = (-1)^{n-1}\frac{x^n(1-\theta)^{n-1}}{(1+\theta x)^n}.$$

Choose $t>0$ such that $-1<-t<x$; then since $0<\theta<1$,

$$0<1-\theta<1+\theta x \quad \text{and} \quad 1+\theta x>1-\theta t>1-t.$$

Hence

$$|R_n| = \frac{|x|^n}{1+\theta x}\left(\frac{1-\theta}{1+\theta x}\right)^{n-1}$$

$$< \frac{|x|^n}{1-t} \to 0 \qquad \text{as } n\to\infty.$$

We have shown, therefore, that

$$\log(1+x) = x - \frac{x^2}{2} + \frac{x^3}{3} - \cdots + (-1)^{n-1}\frac{x^n}{n} + \cdots$$

$$\text{for } -1<x\leqslant 1.$$

In particular, when $x=1$,

$$\log 2 = 1 - \tfrac{1}{2} + \tfrac{1}{3} - \tfrac{1}{4} + \cdots.$$

Theorem As $x\to\infty$, the function $\log x$ increases less rapidly than any power of x. That is $\lim\limits_{x\to\infty} x^{-\alpha}\log x = 0$, for any $\alpha>0$.

Proof. Let $x=e^t$; then

$$x^{-\alpha}\log x = te^{-\alpha t} \to 0 \qquad \text{as } t\to\infty,$$

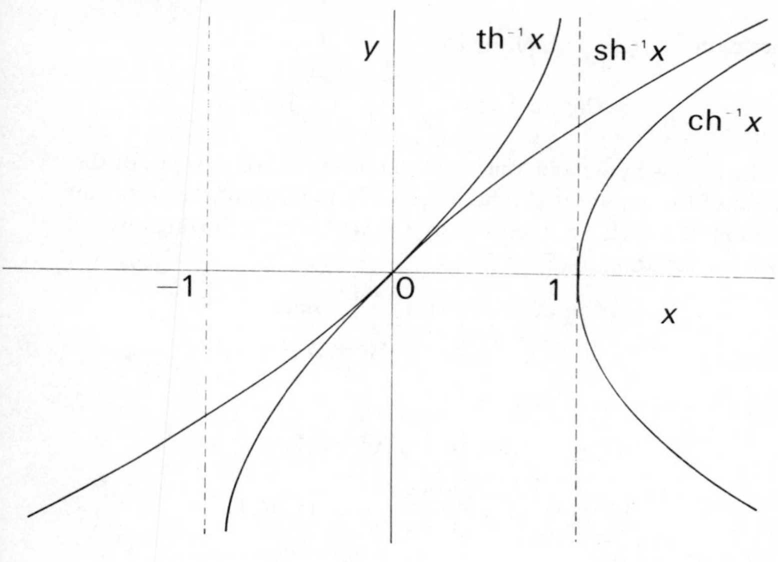

Fig. 6

since $\alpha > 0$ (Section 8.5). Put $x = 1/u$; then the result can be written

$$\lim_{u \to 0} u^{\alpha} \log u = 0, \qquad \alpha > 0.$$

8.7 Inverse hyperbolic functions

The inverse hyperbolic functions are written $\mathrm{ch}^{-1} x$, $\mathrm{sh}^{-1} x$, $\mathrm{th}^{-1} x$; their graphs are shown in Fig. 6.

If $y = \mathrm{sh}^{-1} x$, then $x = \mathrm{sh}\, y$ and

$$\frac{\mathrm{d}y}{\mathrm{d}x} = \frac{1}{\mathrm{ch}\, y} = \frac{1}{\sqrt{(1 + \mathrm{sh}^2 y)}} = \frac{1}{\sqrt{(1 + x^2)}},$$

the positive root being taken because $\mathrm{ch}\, y \geqslant 1$.

Similarly we have

$$D(\mathrm{ch}^{-1} x) = \frac{1}{\sqrt{(x^2 - 1)}}, \qquad x > 1,$$

in this case the negative root is related to the lower branch of the curve; the positive root gives the principal value.

The function $\mathrm{th}^{-1} x$ is defined only for $|x| < 1$;

$$D(\mathrm{th}^{-1} x) = \frac{1}{1 - x^2}.$$

Since $\coth^{-1} x = \text{th}^{-1}(1/x)$, we have

$$D(\coth^{-1} x) = \frac{-1}{x^2-1}, \qquad |x| > 1.$$

The inverse hyperbolic functions can be expressed in terms of the inverse of the exponential function, that is, in terms of the logarithmic function. If $y = \text{sh}^{-1} x$, then $x = \text{sh}\, y = \frac{1}{2}(e^y - e^{-y})$. Solving this quadratic equation in e^y we find

$$e^y = x + \sqrt{(x^2+1)}, \qquad \text{since } e^y > 0.$$

Therefore $$y = \text{sh}^{-1} x = \log[x + \sqrt{(x^2+1)}].$$

Similarly we can show that

$$\text{ch}^{-1} x = \pm \log[x + \sqrt{(x^2-1)}], \qquad x > 1;$$

$$\text{th}^{-1} x = \tfrac{1}{2}\log\left(\frac{1+x}{1-x}\right), \qquad |x| < 1.$$

EXERCISES 8(c)

1. Prove that if $x > -1$, the remainder after the term in x^{n-1} $(n \geqslant 2)$ in the Maclaurin expansion of $(1+x)\log(1+x)$ is numerically less than $|x|^n/(n+1)$.

2. Show that

$$\frac{\frac{1}{2}x^2}{1+x} < x - \log(1+x) < \tfrac{1}{2}x^2 \text{ for } x > 0;$$

show that the inequalities are reversed if $-1 < x < 0$.

3. Show that

$$\log x = (x-a) - \tfrac{1}{2}(x-a)^2 + \tfrac{1}{3}(x-a)^3 - \cdots, \qquad |x-a| < 1$$

4. Show that, if $y > 0$,

$$\log y = 2\left\{\left(\frac{y-1}{y+1}\right) + \frac{1}{3}\left(\frac{y-1}{y+1}\right)^3 + \frac{1}{5}\left(\frac{y-1}{y+1}\right)^5 + \cdots\right\}.$$

5. If $0 < x < 1$, prove that

$$(1+x)\log(1+x) + (1-x)\log(1-x) > 0.$$

Deduce that, if $n > 1$,

$$(n+1)^{1+1/n}(n-1)^{1-1/n} > n^2.$$

6. Show that

$$\int_0^1 \log(1+x)\frac{dx}{x} = 1 - \frac{1}{2^2} + \frac{1}{3^2} - \frac{1}{4^2} + \cdots$$

$$= \frac{1}{2}\left(1 + \frac{1}{2^2} + \frac{1}{3^2} + \cdots\right) = \frac{\pi^2}{12}.$$

7. Prove that if f' exists in $0 < a < x < b$, then there is a number ξ, $a < \xi < b$, such that

$$f(b) - f(a) = \xi f'(\xi) \log(b/a).$$

By taking $f(x) = x^{1/n}$, deduce that

$$\lim_{n\to\infty} n(a^{1/n} - 1) = \log a.$$

8. Prove that, if $x > 1$,

$$\log\left(\frac{x+1}{x}\right) < \frac{1}{x} < \log\left(\frac{x}{x-1}\right).$$

Deduce that $1 + \frac{1}{2} + \frac{1}{3} + \cdots + \frac{1}{n} - \log n$ lies between 1 and $\log(1 + 1/n)$.

9. Evaluate:

(a) $\lim_{x\to 0} \frac{1}{x}\log\left(\frac{xe^x}{e^x - 1}\right)$ (b) $\lim_{x\to 0} \frac{x^2 + \log(1-x^2)}{x^2 - \operatorname{sh}^2 x}$

(c) $\lim_{n\to\infty} \left(\cos\frac{\alpha}{n}\right)^{n^2}$

(d) $\lim_{x\to 0} \frac{\operatorname{th}^{-1} x + \log(1-x) + \operatorname{sh}^2 \frac{1}{2}x^2 + \frac{1}{2}x^2}{x^3(\operatorname{sh} x - x)}$

10. Given $u_n > 0$ and

$$\frac{u_n}{u_{n+1}} = 1 + \frac{1}{n} + \frac{A_n}{n^\lambda},$$

where A_n is bounded and $\lambda > 1$, prove that Σu_n is divergent by taking $d_n = n \log n$ in Kummer's test (Exercise 6(c) No. 2).

8.8 The binomial theorem

$$(1+x)^m = 1 + mx + \frac{m(m-1)}{2!}x^2 + \frac{m(m-1)(m-2)}{3!}x^3 + \cdots$$

$$= 1 + \sum_{n=1}^{\infty} \binom{m}{n} x^n,$$

where

$$\binom{m}{n} = \frac{m(m-1)(m-2)\cdots(m-n+1)}{n!}.$$

The series is absolutely convergent for all m when $|x|<1$; convergent at $x=1$ when $m>-1$ and convergent at $x=-1$ when $m>0$ (Example 4, Section 7.2).

If $f(x)=(1+x)^m$, then

$$f^{(r)}(x) = m(m-1)(m-2)\cdots(m-r+1)(1+x)^{m-r},$$
$$r = 1, 2, 3, \ldots.$$

If m is a positive integer, $f^{(r)}(x)=0$ for $r=m+1, m+2, \ldots$ and the expansion has $m+1$ terms.

Considering the case when m is not a positive integer, then the remainder after n terms,

$$R_n = \frac{x^n}{n!} m(m-1)(m-2)\cdots(m-n+1)(1+\theta x)^{m-n} \qquad (0<\theta<1)$$

$$= \binom{m}{n} x^n (1+\theta x)^{m-n}.$$

Then, if n is sufficiently large to make $m-n<0$,

$$|R_n| < \left|\binom{m}{n}\right| x^n \text{ when } 0<x\leqslant 1,$$

$$\to 0 \text{ as } n\to\infty \qquad \text{whenever the binomial series converges.}$$

For $-1<x<0$, we use Cauchy's form of the remainder,

$$R_n = n\binom{m}{n}\frac{(1-\theta)^{n-1}}{(1+\theta x)^{n-m}} x^n, \qquad 0<\theta<1,$$

$$= m\binom{m-1}{n-1}\left(\frac{1-\theta}{1+\theta x}\right)^{n-1}(1+\theta x)^{m-1}x^n.$$

Since $-1<x<0$ and $0<\theta<1$, then

$$0<1-\theta<1+\theta x, \; 1-|x|<1+\theta x<1+|x|,$$

and $$(1+\theta x)^{m-1} < (1+|x|)^{m-1}, \qquad \text{for } m > 1,$$
$$< (1-|x|)^{m-1}, \qquad \text{for } m < 1.$$

Then
$$|R_n| < M(m, x) \left|\binom{m-1}{n-1}\right| |x|^{n-1},$$

where
$$M(m, x) = m\,|x|\,(1+|x|)^{m-1}, \qquad m > 1$$
$$= m\,|x|\,(1-|x|)^{m-1}, \qquad m < 1.$$

Hence $|R_n| \to 0$ as $n \to \infty$ for $-1 < x < 0$, since the binomial series converges in this interval.

EXERCISES 8(d)

1. Write down the first three terms and the general terms of the expansions in ascending powers of x of the following functions:

   (a) $(1+x)^{-1}$ (b) $(1+x)^{-2}$ (c) $(1+x)^{1/2}$ (d) $(1+x)^{-1/2}$.

2. Find the numerically greatest term or terms in the following expansions:

   (a) $(1+5x)^{-4}$ when $x = \frac{1}{7}$ (b) $(1+x)^{-1/2}$ when $x = \frac{1}{2}$
   (c) $(1-x)^{-3}$ when $x = \frac{4}{5}$.

3. Sum the following series:

   (a) $\dfrac{1}{3} + \dfrac{1.3}{3.6} + \dfrac{1.3.5}{3.6.9} + \dfrac{1.3.5.7}{3.6.9.12} + \cdots$

   (b) $x^2 + \dfrac{1}{3}x^3 + \dfrac{1.3}{3.4}x^4 + \dfrac{1.3.5}{3.4.5}x^5 + \cdots, \; |x| < \dfrac{1}{2}$

   (c) $1 + \dfrac{1.3}{2.4}x^2 + \dfrac{1.3.5.7}{2.4.6.8}x^4 + \cdots, \; |x| < 1.$

4. By expanding $(1+x)^{2n}$ in powers of x prove that
$$1 + \binom{2n}{1} + \binom{2n}{2} + \cdots + \binom{2n}{2n} = 2^{2n},$$
$$1 - \binom{2n}{1} + \binom{2n}{2} - \cdots + \binom{2n}{2n} = 0,$$
   when $n = 1, 2, 3, \ldots$.

Deduce that if $\operatorname{ch} x = \sum_{n=0}^{\infty} \frac{x^{2n}}{(2n)!}$, then $2\operatorname{ch}^2 x = 1 + \operatorname{ch} 2x$.

5. Show that

$$(1+x)^{1+x} = \begin{cases} 1 + x + x^2 + \frac{1}{2}x^3 + \cdots, & \text{when } |x| \text{ is small,} \\ ex^{1+x}\left(1 + \dfrac{1}{2x} - \dfrac{1}{24x^2} + \cdots\right) & \text{when } x \text{ is large.} \end{cases}$$

6. Evaluate:

(a) $\lim_{x \to \infty} [x - \frac{1}{2}(4x^2 + x)^{1/2}]$ (b) $\lim_{x \to \infty} x[e - (1 + \frac{1}{x})^x]$.

8.9 Other methods for finding Maclaurin series

In Example 4, Section 7.9 we proved the uniqueness of the Maclaurin expansion of a function. Consequently we can use the properties of power series developed in Chapter 7 to obtain the Taylor or Maclaurin series of a given function. For example, in Example 3, Section 7.9 we found the Maclaurin series

$$\operatorname{th}^{-1} x = \sum_{n=0}^{\infty} \frac{x^{2n+1}}{2n+1}, \qquad |x| < 1,$$

and in Example 4, Section 7.10 we found the first four non-zero terms of the Maclaurin expansion for $\tan x$,

$$\tan x = x + \frac{1}{3}x^3 + \frac{2}{15}x^5 + \frac{17}{315}x^7 + \cdots,$$

In Section 7.11 we found the series

$$e^{\sin x} = 1 + x + \frac{1}{2}x^2 - \frac{1}{8}x^4 - \frac{1}{15}x^5 + \cdots,$$

by substituting one power series into another. This method could also be used to find the Maclaurin series for $\tan x$.

$$\tan x = \frac{\sin x}{\cos x} = \left(x - \frac{x^3}{3!} + \frac{x^5}{5!} - \frac{x^7}{7!} + \cdots\right) \times \left[\left(\frac{x^2}{2!} - \frac{x^4}{4!} + \frac{x^6}{6!} - \cdots\right)\right]^{-1},$$

then using the binomial theorem

$$\tan x = \left(x - \frac{x^3}{6} + \frac{x^5}{120} - \frac{x^7}{5040} + \cdots\right)\left[1 + \left(\frac{x^2}{2} - \frac{x^4}{24} + \frac{x^6}{720} - \cdots\right) + \left(\frac{x^2}{2} - \frac{x^4}{24} + \cdots\right)^2 + \left(\frac{x^2}{2} - \cdots\right)^3 + \cdots\right]$$

if $|x|$ is sufficiently small. That is

$$\tan x = \left(x - \frac{x^3}{6} + \frac{x^5}{120} - \frac{x^7}{5040} + \cdots\right)\left[1 + \frac{x^2}{2} + \left(\frac{1}{4} - \frac{1}{24}\right)x^4 + \left(\frac{1}{8} - \frac{1}{24} + \frac{1}{720}\right)x^6 + \cdots\right]$$

$$= x + \frac{1}{3}x^3 + \frac{2}{15}x^5 + \frac{17}{315}x^7 + \cdots,$$

as before.

Another method uses Leibniz's formula for the nth derivative of a product (Section 4.4).

Example Let $y = \sin^{-1} x$; then

$$y' = 1/\sqrt{(1 - x^2)} \tag{1}$$

or

$$(1 - x^2)(y')^2 = 1.$$

Differentiating this relation and then cancelling by $2y'$ gives

$$(1 - x^2)y'' - xy' = 0.$$

This is the equation in Example 4, Section 4.4, which was differentiated n times using Leibniz's formula, and led to

$$(1 - x^2)y^{(n+2)} - (2n + 1)xy^{(n+1)} - n^2y^{(n)} = 0. \tag{2}$$

Now put $x = 0$ and write y_n for the value of $y^{(n)}$ at $x = 0$. The successive derivatives at the origin satisfy

$$y_{n+2} - n^2y_n = 0, \qquad n = 1, 2, 3, \ldots.$$

This recurrence relation relates even-order derivatives with y_2 and odd orders with y_1. Since $y_2 = 0$, this implies that

$$y_{2k} = 0, \qquad k = 1, 2, 3, \ldots.$$

For the odd-order derivatives, put $n = 2k - 1$, then

$$\begin{aligned} y_{2k+1} &= (2k-1)^2 y_{2k-1} \qquad (k = 1, 2, 3, \ldots) \\ &= (2k-1)^2(2k-3)^2 y_{2k-3} \\ &\quad \vdots \\ &= (2k-1)^2(2k-3)^2(2k-5)^2 \cdots 5^2.3^2.1^2 y_1 \\ &= 1^2.3^2.5^2 \cdots (2k-1)^2, \qquad k = 1, 2, 3, \ldots. \end{aligned}$$

Whence the Maclaurin series for y is given by

$$\begin{aligned} y &= \sum_{n=1}^{\infty} y_n \frac{x^n}{n!} \\ &= \sum_{k=1}^{\infty} \frac{1^2.3^2.5^2 \cdots (2k-1)^2}{(2k-1)!} x^{2k-1} \\ &= x + \frac{1}{2}\frac{x^3}{3} + \frac{1.3}{2.4}\frac{x^5}{5} + \frac{1.3.5}{2.4.6}\frac{x^7}{7} + \cdots \end{aligned}$$

for sufficiently small $|x|$.

Since we cannot solve eq. (2) we cannot find the general value of $y^{(n)}$ and so cannot study the behaviour of the remainder. Of course, in this particular example we can also find the Maclaurin series from eq. (1) by integrating the binomial expansion of the right-hand side of the equation.

EXERCISES 8(e)

1. Obtain, for sufficiently small $|x|$, the following expansions:

(a) $\dfrac{2e^x}{e^x + 1} = 1 + \dfrac{1}{2}x - \dfrac{1}{24}x^3 + \cdots$

(b) $\dfrac{[(1 + 3x^2)e^x]^{1/2}}{1 - x} = 1 + \dfrac{3}{2}x + \dfrac{25}{8}x^2 + \cdots$

(c) $e^{x \sin x} = 1 + x^2 + \frac{1}{3}x^4 + \frac{1}{120}x^6 + \cdots$

(d) $\log(\cos x + \sin x) = x - x^2 + \frac{2}{3}x^3 - \frac{2}{3}x^4 + \cdots$

(e) $\log \sin x = \log x - \frac{1}{6}x^2 - \frac{1}{180}x^4 - \cdots$

(f) $\sin^{-1} x = x + \dfrac{1}{2}\dfrac{x^3}{3} + \dfrac{1.3}{2.4}\dfrac{x^5}{5} + \dfrac{1.3.5}{2.4.6}\dfrac{x^7}{7} + \cdots$

(g) $\text{th}\, x = x - \frac{1}{3}x^3 + \frac{2}{15}x^5 - \frac{17}{315}x^7 + \cdots$

(h) $\text{sh}^{-1} x = x - \frac{1}{2}\frac{x^3}{3} + \frac{1.3}{2.4}\frac{x^5}{5} - \frac{1.3.5}{2.4.6}\frac{x^7}{7} + \cdots$

(i) $\tan^{-1}(\text{sh}\, x) = x - \frac{1}{6}x^3 + \frac{1}{24}x^5 - \cdots$

(j) $$e^{ax}\cos bx = 1 + ax + \tfrac{1}{2}(a^2 - b^2)x^2 + \tfrac{1}{6}(a^3 - 3ab^2)x^3 + (a^4 - 6a^2b^2 + b^4)\frac{x^4}{4!} + \cdots.$$

2. Show that

$$e^{\tan x} = 1 + x + \tfrac{1}{2}x^2 + \tfrac{1}{2}x^3 + \tfrac{3}{8}x^4 + \tfrac{37}{120}x^5 + \cdots.$$

Hence obtain expansions as far as the term in x^4 of

(a) $\sec^2 x \exp(\tan x)$ (b) $\text{ch}(\tan x)$.

3. Show that if $\log y = \tan^{-1} x$, where the inverse tangent has that value between $\pm\frac{1}{2}\pi$, then $(1 + x^2)y' = y$ and

$$(1 + x^2)y^{(n+1)} + (2nx - 1)y^{(n)} + n(n-1)y^{(n-1)} = 0, \qquad n \geqslant 2.$$

Deduce that

$$y = 1 + x + \tfrac{1}{2}x^2 - \tfrac{1}{6}x^3 - \tfrac{7}{24}x^4 + \tfrac{1}{24}x^5 + \cdots.$$

Evaluate

$$\lim_{x\to 0}\{e^x(1 - \tfrac{1}{3}x^3) - \exp(\tan^{-1} x)\}x^{-5}.$$

4. Show that if $y = \sin(m \sin^{-1} x)$, then

$$(1 - x^2)y'' - xy' + m^2 y = 0.$$

Differentiate this equation n times using Leibniz's theorem and hence obtain the Maclaurin expansion

$$y = mx + \frac{m(1^2 - m^2)}{3!}x^3 + \frac{m(1^2 - m^2)(3^2 - m^2)}{5!}x^5 + \cdots.$$

5. Given $f(x) = \cos[\pi\sqrt{(1 + x)}]$, show that

$$4(1 + x)f''(x) + 2f'(x) + \pi^2 f(x) = 0$$

and

$$4f^{(n+2)}(0) + 2(2n + 1)f^{(n+1)}(0) + \pi^2 f^{(n)}(0) = 0, \qquad n \geqslant 1.$$

Hence find the Maclaurin expansion of $f(x)$ as far as the term in x^4.

6. Given $f(x) = \int_1^x t^t \, dt$, show that $f''(x) = (1 + \log x) f'(x)$. Hence obtain the Taylor expansion

$$f(x) = (x-1) + \tfrac{1}{2}(x-1)^2 + \tfrac{1}{3}(x-1)^3 + \tfrac{1}{8}(x-1)^4 + \cdots.$$

7. Evaluate:

(a) $\lim_{x \to 0} \dfrac{\sin x \sin^{-1} x - x^2}{x^6}$ (b) $\lim_{x \to 0} \dfrac{x^2 e^{-x^2} - \sin^2 x}{\log(1 + x^2) - x \sin x}$

(c) $\lim_{x \to 0} \dfrac{\log(1 + 2x) - 2x + 2x^2}{x^2 \tan^{-1} x}$

(d) $\lim_{x \to 0} \dfrac{e^{-x}\sqrt{(1 + x)} - \cos \sqrt{x}}{x \log(1 + x)}$

(e) $\lim_{x \to 0} \dfrac{\operatorname{ch} x \log(1 + x^2) - x^2}{x^6}$

(f) $\lim_{x \to 1} \dfrac{\tan(\log x) - \log(\tan \frac{1}{4}\pi x)}{\log x}$

(g) $\lim_{x \to 0} \dfrac{e^{\sin x} - e^x}{x^3}$ (h) $\lim_{x \to 0} \left(\dfrac{\sin x}{x}\right)^{1/x^2}$

(i) $\lim_{x \to 0} \dfrac{(\sin x)^{\sin x} - x^x}{(\sin x)^x - x^{\sin x}}$.

8. Find the power series for $(1 + x)^{-1} \log(1 + x)$ when $|x| < 1$. Deduce that, when $|x| < 1$,

$$\left[\log(1 + x)\right]^2 = 2 \sum_{n=1}^{\infty} (-1)^{n-1}\left(1 + \frac{1}{2} + \frac{1}{3} + \cdots + \frac{1}{n}\right)\frac{x^{n+1}}{n+1}.$$

CHAPTER 9

The Riemann Integral

9.1 Upper and lower sums

In elementary work the existence of the definite integral as the limit of a sum is made obvious by an appeal to geometrical argument. This has some disadvantages; it always uses the graph of $y = f(x)$ assumed continuous and it presupposes the concepts of curve and area. The first rigorous theory of integration, in which the graphical discussion was replaced by arithmetical argument, was given by G.F. Riemann in about 1854.

Suppose f is a function defined and bounded in a closed interval $[a, b]$. Divide the interval by any finite set of points

$$a = x_0 < x_1 < x_2 < \cdots < x_r < \cdots < x_{n-1} < x_n = b.$$

This mode of subdivision is called a *net* $\mathfrak{N}$, and divides $[a, b]$ into n subintervals $[x_{r-1}, x_r]$ of lengths

$$\delta_r = x_r - x_{r-1}, \qquad r = 1, 2, \ldots, n.$$

The longest of the subintervals is called the *norm* of the net and is denoted by

$$\|\mathfrak{N}\| = \mu = \max_{1 \leqslant r \leqslant n} \delta_r.$$

Let the supremum and infimum of f on $[x_{r-1}, x_r]$ be M_r, m_r respectively, and form the sums

$$s(\mathfrak{N}) = \sum_{r=1}^{n} m_r(x_r - x_{r-1}), \quad S(\mathfrak{N}) = \sum_{r=1}^{n} M_r(x_r - x_{r-1}).$$

These sums are called the *lower* and *upper sums* corresponding to the net $\mathfrak{N}$.

Example 1 If $f(x) = (1+x)^{-1}, 0 \leqslant x \leqslant 1$, then taking $x_r = r/n$ we have a net $\mathfrak{N}$ in which $\delta_r = 1/n, r = 1, 2, \ldots, n$. In the subinterval $[x_{r-1}, x_r]$, $m_r = (1+x_r)^{-1}$ and $M_r = (1+x_{r-1})^{-1}$. Then we have the sums

$$s(\mathfrak{N}) = \sum_{r=1}^{n} \frac{1}{1+r/n} \frac{1}{n} = \sum_{r=1}^{n} \frac{1}{n+r},$$

and

$$S(\mathfrak{N}) = \sum_{r=1}^{n} \frac{1}{n+r-1}.$$

Example 2 Let $f(x) = x^2, 0 < a \leqslant x \leqslant b$, and construct the net $\mathfrak{N}$ by making the points of subdivision form a geometric progression. That is, take $x_r = a\rho^r$, where $x_n = b = a\rho^n$. Then

$$\begin{aligned} S(\mathfrak{N}) &= \sum_{r=1}^{n} x_r^2(x_r - x_{r-1}) = a^3(\rho-1)\sum_{r=1}^{n} \rho^{3r-1} \\ &= a^3(\rho-1)\rho^2(1-\rho^{3n})/(1-\rho^3). \end{aligned}$$

Thus $$S(\mathfrak{N}) = \frac{(b^3-a^3)\rho^2}{1+\rho+\rho^2}, \quad \text{since } a\rho^n = b.$$

Similarly $$s(\mathfrak{N}) = \frac{b^3-a^3}{1+\rho+\rho^2}.$$

If on $[a, b]$, $\sup f = M$ and $\inf f = m$, then

$$m \leqslant m_r \leqslant M_r \leqslant M$$

for each r, and

$$m\delta_r \leqslant m_r\delta_r \leqslant M_r\delta_r \leqslant M\delta_r.$$

Then summing for $r = 1, 2, 3, \ldots, n$

$$m(b-a) \leqslant s(\mathfrak{N}) \leqslant S(\mathfrak{N}) \leqslant M(b-a).$$

This shows that for any net $\mathfrak{N}$ the lower sum cannot exceed the corresponding upper sum. This must be generalised to show that *no* lower sum exceeds *any* upper sum.

Lemma If, to a given net $\mathfrak{N}$, extra dividing points are added, the upper sum is not increased and the lower sum is not decreased.

Proof. Suppose there is only one extra point ξ and that it falls within (x_{k-1}, x_k). Then the new net $\mathfrak{N}_1$ coincides with $\mathfrak{N}$ for each subinterval except $[x_{k-1}, x_k]$, which is replaced by two subintervals $[x_{k-1}, \xi]$ and

$\xi, x_k]$. Let the suprema of f on these subintervals be M'_k and M''_k respect-vely.

Considering the upper sums we have

$$S(\mathfrak{N}) - S(\mathfrak{N}_1) = M_k(x_k - x_{k-1}) - [M'_k(\xi - x_{k-1}) + M''_k(x_k - \xi)].$$

Iow $M'_k \leqslant M_k$ and $M''_k \leqslant M_k$, and so

$$S(\mathfrak{N}) - S(\mathfrak{N}_1) \geqslant M_k(x_k - x_{k-1} - \xi + x_{k-1} - x_k + \xi) = 0.$$

'hat is $S(\mathfrak{N}_1) \leqslant S(\mathfrak{N})$.

Now suppose the lemma is true for a net $\mathfrak{N}_p$ formed from $\mathfrak{N}$ by adding extra points. Then $S(\mathfrak{N}_{p+1}) \leqslant S(\mathfrak{N}_p)$ since $\mathfrak{N}_{p+1}$ is obtained from $\mathfrak{N}_p$ by dding one extra point. We have just proved the case $p = 1$; the general ase follows by induction.

The corresponding result for lower sums,

$$s(\mathfrak{N}_1) \geqslant s(\mathfrak{N}),$$

left as an exercise.

We are now in a position to prove the following.

'heorem No lower sum exceeds any upper sum; no upper sum is less han any lower sum.

roof. Let $\mathfrak{N}_1$, $\mathfrak{N}_2$ be two distinct nets with lower and upper sums ndicated by an appropriate subscript. Form a net $\mathfrak{N}_3$ by superimposing $\mathfrak{N}_1$ and $\mathfrak{N}_2$. Since $\mathfrak{N}_3$ is formed by adding points to $\mathfrak{N}_1$

$$s_3(\mathfrak{N}_3) \geqslant s_1(\mathfrak{N}_1).$$

ut we can also regard $\mathfrak{N}_3$ as being formed by adding points to $\mathfrak{N}_2$, thus

$$S_3(\mathfrak{N}_3) \leqslant S_2(\mathfrak{N}_2).$$

n any case $s_3(\mathfrak{N}_3) \leqslant S_3(\mathfrak{N}_3)$; that is

$$s_1(\mathfrak{N}_1) \leqslant s_3(\mathfrak{N}_3) \leqslant S_3(\mathfrak{N}_3) \leqslant S_2(\mathfrak{N}_2).$$

herefore

$$s_1(\mathfrak{N}_1) \leqslant S_2(\mathfrak{N}_2).$$

imilarly we can show that

$$s_2(\mathfrak{N}_2) \leqslant S_1(\mathfrak{N}_1).$$

hat is no lower sum exceeds any upper sum, and so we can write the ower and upper sums simply as s, S without specifying the net to which ey correspond.

9.2 Upper and lower integrals

We have seen that the lower and upper sums corresponding to a net $\mathfrak{N}$ must satisfy the inequalities

$$m(b-a) \leqslant s(\mathfrak{N}) \leqslant S(\mathfrak{N}) \leqslant M(b-a).$$

The theorem we have just proved shows that this can be replaced by

$$m(b-a) \leqslant s \leqslant S \leqslant M(b-a),$$

where s, S are now any lower and upper sums, not necessarily corresponding to the same net. This shows that these sums are bounded; in particular the upper sums must have an infimum and the lower sums a supremum.

We define, for all possible nets,

$$I = \sup s \quad \text{and} \quad J = \inf S,$$

and call them the *lower* and *upper integrals* of f over (a, b). We can write

$$I = \underline{\int_a^b} f(x)\,\mathrm{d}x, \quad J = \overline{\int_a^b} f(x)\,\mathrm{d}x.$$

Example Let

$$f(x) = \begin{cases} 0, & x \text{ irrational}, \\ 1, & x \text{ rational}; \end{cases}$$

and let the interval be $[a, b]$.

Then for any net $\mathfrak{N}, m_r = 0, M_r = 1$ and so

$$s = 0, \quad S = \sum_{r=1}^{n} (x_r - x_{r-1}) = b - a.$$

Thus $$I = 0 \quad \text{and} \quad J = b - a.$$

Since the infimum and supremum of a bounded function always exist, a function bounded in a finite interval must possess lower and upper integrals. We now show that it is possible to find nets for which the upper sums are arbitrarily close to the upper integral.

Darboux's theorem $\forall \epsilon > 0; \exists\, \theta > 0$ such that

$$S - J < \epsilon \quad \text{and} \quad I - s < \epsilon$$

for all nets of norm $\leqslant \theta$.

Proof. Consider the upper sums S. Since $J = \inf S$, there is a net $\mathfrak{N}_1$ with upper sum S_1 such that

$$S_1 < J + \tfrac{1}{2}\epsilon \qquad (1)$$

for any given $\epsilon > 0$.

Let this net be $a = a_0 < a_1 < a_2 < \cdots < a_{p-1} < a_p = b$, and choose θ so that $\theta < \min\limits_{1 \leqslant r \leqslant p} (a_r - a_{r-1})$.

Now let $\mathfrak{N}_2$ be any net of norm $\mu \leqslant \theta$; thus

$$\mathfrak{N} : a = x_0 < x_1 < x_2 < \cdots < x_{n-1} < x_n = b$$

and

$$\mu = \max_{1 \leqslant r \leqslant n} (x_r - x_{r-1}) \leqslant \theta.$$

Superimpose nets $\mathfrak{N}_1$ and $\mathfrak{N}_2$ to form a net $\mathfrak{N}_3$ with upper sum S_3. By our choice of nets no two points $a_1, a_2, \ldots, a_p$ can lie in the same subinterval of $\mathfrak{N}_2$. For the longest subinterval $[x_{r-1}, x_r]$ is less than the shortest subinterval $[a_{r-1}, a_r]$, and so each subinterval $[x_{r-1}, x_r]$ of $\mathfrak{N}_2$ contains at most one point of $\mathfrak{N}_1$. Consider the difference $S_2 - S_3$; when the subinterval $[x_{r-1}, x_r]$ contains no point of $\mathfrak{N}_1$ or when the a_k coincides with an end point of the subinterval, S_2 and S_3 both contain the same terms. Consequently $S_2 - S_3$ is made up of terms for which there is an a_k in the subinterval $[x_{r-1}, x_r]$. Thus, if M_r, M_r', M_r'' are the suprema of f on $[x_{r-1}, x_r]$, $[x_{r-1}, a_k]$ and $[a_k, x_r]$ respectively,

$$S_2 - S_3 = \sum \{M_r(x_r - x_{r-1}) - M_r'(a_k - x_{r-1}) - M_r''(x_r - a_k)\},$$

where the summation is taken over all the subintervals of $\mathfrak{N}_2$ which contain a point a_k of $\mathfrak{N}_1$. Rearranging the terms we have

$$\begin{aligned} S_2 - S_3 &= \sum \{(M_r - M_r')(a_k - x_{r-1}) + (M_r - M_r'')(x_r - a_k)\} \\ &\leqslant 2H \sum (a_k - x_{r-1} + x_r - a_k) = 2H \sum (x_r - x_{r-1}), \end{aligned}$$

where $H = \sup |f|$ on $[a, b]$. Thus

$$S_2 - S_3 \leqslant 2H(p-1)\mu \leqslant 2H(p-1)\theta. \qquad (2)$$

Also, since $\mathfrak{N}_3$ is formed by adding points to $\mathfrak{N}_1$,

$$\begin{aligned} S_3 &\leqslant S_1 \\ &< J + \tfrac{1}{2}\epsilon, \qquad \text{from (1)}. \end{aligned}$$

Then, using (2),

$$S_2 < J + 2H(p-1)\theta + \tfrac{1}{2}\epsilon.$$

Therefore, by choosing $\theta < \epsilon/4H(p-1)$, we have

$$S_2 < J + \epsilon.$$

Since $\mathfrak{N}_2$ is any net of norm $\mu \leqslant \theta$, we have established the theorem for upper sums. The argument for lower sums is similar and is left as an exercise.

When the upper and lower integrals of a bounded function are equal, their common value is called the *Riemann integral* of the function over the given integral. We write this integral in the usual way

$$\int_a^b f(x)\,dx.$$

When this integral exists we say that the function is *integrable-R* over the interval (a, b).

9.3 Condition for integrability

Theorem A necessary and sufficient condition for a function f bounded on $[a, b]$, to be integrable-R over (a, b) is that, corresponding to any $\epsilon > 0$, there exists a net for which $S - s < \epsilon$.

Proof. First suppose f is integrable-R over (a, b); that is, suppose $I = J$. From Darboux's theorem (Section 9.2) we can find a net $\mathfrak{N}$ of norm μ such that, for any $\epsilon > 0$,

$$S - J < \tfrac{1}{2}\epsilon \quad \text{and} \quad I - s < \tfrac{1}{2}\epsilon.$$

Then using $I = J$, we have

$$S - s = (S - J) + (I - s) < \epsilon,$$

which shows the condition is necessary.

To show the condition is sufficient assume that for a net $\mathfrak{N}$ of norm μ, $S - s < \epsilon$. Now suppose $I < J$ (we know that for any net $I \leqslant J$) and let $J - I = \lambda > 0$. Then

$$\begin{aligned} \lambda &\leqslant S - s \qquad \text{(because } s \leqslant I \leqslant J \leqslant S\text{)} \\ &< \epsilon, \end{aligned}$$

and since ϵ is arbitrary this implies $\lambda = 0$. That is $I = J$ and f is integrable-R over (a, b).

Notice that this condition for integrability together with Darboux's theorem implies that if f is integrable-R over (a, b) and if $\lim_{n \to \infty} s$ (or $\lim_{n \to \infty} S$) exists, then this limit is the value of the integral.

Example 1 $f(x) = (1+x)^{-1}, 0 \leqslant x \leqslant 1.$

From Example 1, Section 9.1, we have

$$s = \sum_{r=1}^{n} \frac{1}{n+r} \quad \text{and} \quad S = \sum_{r=1}^{n} \frac{1}{n-1+r}.$$

Then
$$S - s = \frac{1}{2n} < \epsilon \qquad \text{for } n > N = [1/2\epsilon].$$

That is, for nets of norm less than $1/N$, $S - s < \epsilon$ and so f is integrable-R over $(0, 1)$.

We shall see later (Example 3, Section 9.10) that

$$\lim_{n \to \infty} s = \lim_{n \to \infty} \sum_{r=1}^{n} \frac{1}{n+r} = \log 2.$$

Alternatively we can say that, since the function is integrable

$$\lim_{n \to \infty} s = \int_0^1 \frac{dx}{1+x}.$$

Example 2 $f(x) = x, a \leqslant x \leqslant b.$

Let the net

$$\mathfrak{N} : a = x_0 < x_1 < x_2 < \cdots < x_n = b$$

be formed by taking $x_r = a + rh$, where $x_0 = a$ and $x_n = b = a + nh$; that is $h = (b-a)/n$. Then

$$s = \sum_{r=1}^{n} x_{r-1}(x_r - x_{r-1}) \quad \text{and} \quad S = \sum_{r=1}^{n} x_r(x_r - x_{r-1}).$$

Writing $x_r = a + rh$, we have

$$\begin{aligned} S &= h \sum_{r=1}^{n} (a + rh) \\ &= ah \sum_{r=1}^{n} 1 + h^2 \sum_{r=1}^{n} r \\ &= a.nh + \tfrac{1}{2}n(n+1)h^2 \\ &= a(b-a) + \tfrac{1}{2}(b-a)(b-a+h) \\ &= \tfrac{1}{2}(b^2 - a^2) + \tfrac{1}{2}h(b-a). \end{aligned}$$

Similarly $$s = \tfrac{1}{2}(b^2 - a^2) - \tfrac{1}{2}h(b-a).$$

Then $$\begin{aligned} S - s &= h(b-a) \\ &< \epsilon \qquad \text{for } h < \epsilon/(b-a), \end{aligned}$$

that is for $n > (b-a)^2/\epsilon$. Therefore the function is integrable-R over (a, b) and

$$\int_a^b x \, dx = \lim_{n \to \infty} S = \tfrac{1}{2}(b^2 - a^2).$$

Example 3 $f(x) = x^2, 0 < a \leqslant x \leqslant b.$

In Example 2, Section 9.1, we took $x_r = a\rho^r$, where $x_n = b = a\rho^n$ and found

$$s = \frac{b^3 - a^3}{1 + \rho + \rho^2}.$$

Now $\rho = (b/a)^{1/n} \to 1$ as $n \to \infty$ (See Exercises 5(a), No. 4). Therefore

$$\lim_{n \to \infty} s = \frac{1}{3}(b^3 - a^3) = \int_a^b x^2 \, dx.$$

9.4 Monotonic and continuous functions

Theorem If f is bounded and monotonic in $[a, b]$, then it is integrable-R over (a, b).

Proof. Suppose f is monotonic increasing. Then for any net $\mathfrak{N}$ of norm μ, $m_r = f(x_{r-1})$, $M_r = f(x_r)$ and

$$\begin{aligned} S - s &= \sum_{r=1}^{n} [f(x_r) - f(x_{r-1})](x_r - x_{r-1}) \\ &< \mu \sum_{r=1}^{n} [f(x_r) - f(x_{r-1})] = \mu[f(b) - f(a)], \\ &< \epsilon \qquad \text{choosing } \mu < \epsilon/[f(b) - f(a)]. \end{aligned}$$

The proof for decreasing functions is very similar and is left as an exercise.

Theorem If f is continuous in $[a, b]$, then it is integrable-R over (a, b).

Proof. Since f is continuous in $[a, b]$ a net $\mathfrak{N}$ can be found such that $M_r - m_r < k\epsilon$ $(k > 0)$ in each $[x_{r-1}, x_r]$. Then

$$S - s = \sum_{r=1}^{n} (M_r - m_r)(x_r - x_{r-1})$$

$$< k\epsilon \sum_{r=1}^{n} (x_r - x_{r-1}) = k\epsilon(b-a).$$

herefore $S - s < \epsilon$ if $k < 1/(b-a)$ and so the continuous function f is tegrable-R over (a, b).

XERCISES 9(a)

1. Find lower and upper sums for the following functions over the given intervals:

   (a) $f(x) = k$ (constant), $a \leqslant x \leqslant b$
   (b) $f(x) = x^2, 0 \leqslant x \leqslant a$
   (c) $f(x) = (1 + x^2)^{-1}, 0 \leqslant x \leqslant 1$
   (d) $f(x) = x^{-1}, 0 < a \leqslant x \leqslant b$
   (e) $f(x) = [x], 0 \leqslant x \leqslant 4$
   (f) $f(x) = e^x, a \leqslant x \leqslant b$
   (g) $f(x) = \sin x, 0 \leqslant x \leqslant \frac{1}{2}\pi$.

2. From the results of Question 1 above deduce the value of $\int_a^b f(x)\,dx$.

3. By interpreting the sum as an upper or lower sum of a suitable function over a suitable interval, evaluate the following limits:

   (a) $\lim_{n \to \infty} \frac{1}{n^2} \sum_{r=1}^{n} [r(n-r)]^{1/2}$

   (b) $\lim_{n \to \infty} \sum_{r=1}^{n} (n^2 + r)^{-1}$

   (c) $\lim_{n \to \infty} n^5 \sum_{r=1}^{n} (n^2 + r^2)^{-3}$.

4. Form the upper sum for the function $(1 + x^2)^{-2}$ over $0 \leqslant x \leqslant 1$ by dividing the interval into n subdivisions (a) of equal length and (b) of lengths proportional to $1, 2, \ldots, n$. Hence show that

$$\lim_{n \to \infty} n^3 \sum_{r=0}^{n-1} (n^2 + r^2)^{-2} = (\pi + 2)/8.$$

   Find the value of

$$\lim_{n \to \infty} n^3 (n+1)^3 \sum_{r=0}^{n-1} \frac{r+1}{[n^2(n+1)^2 + r^2(r+1)^2]^2}.$$

9.5 The integral as a limit of a sum

Let f be a function defined and bounded in a finite interval $[a, b]$. Taking any net

$$\mathfrak{N} : a = x_0 < x_1 < x_2 < \cdots < x_{n-1} < x_n = b,$$

of norm μ, we choose in each $[x_{r-1}, x_r]$ any point ξ_r, $x_{r-1} \leqslant \xi_r \leqslant x_r$, and form the sum

$$\sigma = \sum_{r=1}^{n} f(\xi_r)(x_r - x_{r-1}).$$

Clearly, for any net $\mathfrak{N}$, $s \leqslant \sigma \leqslant S$.

We now show that

$$\lim_{\|\mathfrak{N}\| \to 0} \sum_{\mathfrak{N}} f(\xi_r)(x_r - x_{r-1}) = \int_a^b f(x)\,\mathrm{d}x$$

in the sense that if one side of the equation exists then so does the other.

Theorem If $\int_a^b f(x)\,\mathrm{d}x$ exists, then given $\epsilon > 0$, we can find $\mu > 0$ such that

$$\left| \sigma - \int_a^b f(x)\,\mathrm{d}x \right| < \epsilon,$$

for all nets of norm $< \mu$, and for all points $\xi_1, \xi_2, \ldots, \xi_n$ such that $x_{r-1} \leqslant \xi_r \leqslant x_r$, $r = 1, 2, \ldots, n$.

Proof. For any net $\mathfrak{N}$ and for any choice of ξ_r in $[x_{r-1}, x_r]$, $r = 1, 2, \ldots, n$, we have $s \leqslant \sigma = \sum_{r=1}^{n} f(\xi_r)(x_r - x_{r-1}) \leqslant S$. Since f is integrable $I = J = \int_a^b f(x)\,\mathrm{d}x$, and

$$s \leqslant \int_a^b f(x)\,\mathrm{d}x \leqslant S;$$

for all $\epsilon > 0$ there exists a net of norm μ, such that $S - s < \epsilon$. Then

$$\begin{aligned} \left| \sigma - \int_a^b f(x)\,\mathrm{d}x \right| &\leqslant S - s \\ &< \epsilon, \end{aligned}$$

for nets of norm $< \mu$ by Darboux's theorem.

Theorem Let f be a function bounded in $[a, b]$. Suppose a fixed numb λ exists such that, to every $\epsilon > 0$ there exists a $\mu > 0$ such that $|\sigma - \lambda| <$ for all nets of norm $< \mu$ and all points ξ_r, $x_{r-1} \leqslant \xi_r \leqslant x_r$, $r = 1, 2, \ldots, n$. Then f is integrable-R over (a, b) and $\lambda = \int_a^b f(x)\,\mathrm{d}x$.

Let I, J be the lower and upper integrals of f over (a, b) and let $\mathfrak{N}$ be a net of norm μ_1 such that

$$J - S < \tfrac{1}{3}\epsilon.$$

In each subinterval $[x_{r-1}, x_r]$, choose ξ_r such that

$$M_r - f(\xi_r) < \frac{\epsilon}{3(b-a)}.$$

Then
$$|S - \sigma| \leqslant \sum_{r=1}^{n} |M_r - f(\xi_r)|(x_r - x_{r-1})$$
$$< \tfrac{1}{3}\epsilon.$$

Also, by hypothesis, $|\sigma - \lambda| < \frac{1}{3}\epsilon$ for nets of norm $< \mu_2$. Hence

$$|J - \lambda| \leqslant |J - S| + |S - \sigma| + |\sigma - \lambda|$$
$$< \epsilon, \qquad \text{for nets of norm } < \mu = \min(\mu_1, \mu_2).$$

Therefore $J = \lambda$. Similarly it can be proved that $I = \lambda$, and so

$$I = J = \lambda = \int_a^b f(x)\,dx.$$

EXERCISES 9(b)

1. Show that if f is bounded and integrable-R over (a, b), then for any $\epsilon > 0$,

$$\left| \sum_{r=1}^{n} [f(\xi_r) - f(\xi_r')]\,(x_r - x_{r-1}) \right| < \epsilon,$$

for every net of norm $< \mu$ and every pair of choices of the ξ's in that net.

2. Show that if $|\sum_{r=1}^{n} [f(\xi_r) - f(\xi_r')]\,(x_r - x_{r-1})| < \epsilon$ for every pair of choices of the ξ's, then

$$\sum_{r=1}^{n} |f(\xi_r) - f(\xi_r')|(x_r - x_{r-1}) < \epsilon.$$

3. Prove that a bounded function f is integrable-R over (a, b) if and only if for every $\epsilon > 0$ we can find a $\mu > 0$ such that

$$\sum_{r=1}^{n} |f(\xi_r) - f(\xi_r')|(x_r - x_{r-1}) < \epsilon,$$

for every net of norm $< \mu$ and every choice of ξ_r, ξ_r' in $[x_{r-1}, x_r]$.

4. A function $\alpha(x)$, defined for $[a, b]$, is called a *step function* if there is a net

$$\mathfrak{N} : a = x_0 < x_1 < x_2 < \cdots < x_{n-1} < x_n = b$$

such that $\alpha(x) = \alpha_r$ (constant) for $x_{r-1} < x < x_r$ $(r = 1, 2, \ldots, n)$. Prove that a step function is integrable-R over (a, b) and

$$\int_a^b \alpha(x)\,\mathrm{d}x = \sum_{r=1}^{n} \alpha_r(x_r - x_{r-1}).$$

5. A function f is defined and bounded on $[a, b]$. Prove that f is integrable-R over (a, b) if and only if for every $\epsilon > 0$ there exist step functions $\alpha(x)$, $\beta(x)$ such that $\alpha(x) \leqslant f(x) \leqslant \beta(x)$ for $a \leqslant x \leqslant b$ and

$$\int_a^b [\beta(x) - \alpha(x)]\,\mathrm{d}x < \epsilon.$$

6. Show that if f is defined and bounded on $[a, b]$ and $f(x) = k$ (constant) for $a < x < b$, then f is integrable-R over (a, b) and

$$\int_a^b f(x)\,\mathrm{d}x = k(b - a),$$

whatever the values of $f(a)$ and $f(b)$.

7. Show that if f is defined but not bounded on $[a, b]$, then f cannot be integrable-R over (a, b).

9.6 Properties of a definite integral

In the definition of integral given above we assumed that $a < b$. If $b < a$ we define

$$\int_b^a f(x)\,\mathrm{d}x = -\int_a^b f(x)\,\mathrm{d}x,$$

whenever f is integrable-R over (a, b). Furthermore, we define

$$\int_a^a f(x)\,\mathrm{d}x = 0.$$

Property 1 If f and g are both integrable-R over (a, b), then so is $f + g$ and $\int_a^b [f(x) + g(x)]\,\mathrm{d}x = \int_a^b f(x)\,\mathrm{d}x + \int_a^b g(x)\,\mathrm{d}x$.

Proof. Let $S, S', \bar{S}$ be the upper sums of f, g and $f + g$ respectively for a net $\mathfrak{N}$; let $s, s', \bar{s}$ be the corresponding lower sums. Writing the suprema f $f, g, f+g$ in $[x_{r-1}, x_r]$ as $M_r, M'_r, \bar{M}_r$ and the infima as m_r, m'_r and $\bar{m}_r$, ve have

$$\bar{M}_r \leqslant M_r + M'_r \quad \text{and} \quad \bar{m}_r \geqslant m_r + m'_r \qquad \text{(see Exercises 2(b) No. 3).}$$

lence
$$\bar{S} \leqslant S + S' \quad \text{and} \quad \bar{s} \geqslant s + s',$$

nd
$$\bar{J} \leqslant J + J' \quad \text{and} \quad \bar{I} \geqslant I + I',$$

here the I's and J's are the respective lower and upper integrals. For any ounded function $\bar{I} \leqslant \bar{J}$; thus

$$I + I' \leqslant \bar{I} \leqslant \bar{J} \leqslant J + J'.$$

ut f and g are integrable-R over (a, b), and

$$I = J, \quad I' = J'.$$

herefore $\bar{I} = \bar{J}$ and the result follows.

roperty 2 If f is integrable-R over (a, b), so also is f^2.

roof. Since f is integrable, for any $\epsilon > 0$, there is a $\mu > 0$ such that, for ny net of norm $< \mu$,

$$\sum_{r=1}^{n} (M_r - m_r)(x_r - x_{r-1}) < \epsilon,$$

here m_r, M_r are the infimum and supremum of f in $[x_{r-1}, x_r]$. Let m'_r, M'_r be the infimum and supremum of f^2 in $[x_{r-1}, x_r]$, and let $\sup |f| = K$ $[a, b]$. Then

$$M'_r - m'_r \leqslant 2K(M_r - m_r).$$

ence

$$\sum_{r=1}^{n} (M'_r - m'_r)(x_r - x_{r-1}) \leqslant 2K \sum_{r=1}^{n} (M_r - m_r)(x_r - x_{r-1})$$
$$< 2K\epsilon.$$

ince $4fg = (f+g)^2 - (f-g)^2$ this result gives the following.

roperty 2′ If f and g are both integrable-R over (a, b), then so is the roduct fg.

roperty 3 If f is integrable-R over (a, b) and $f \geqslant 0$ in $[a, b]$, then $\int_a^b f(x)\,dx \geqslant 0$.

roof. Since $f \geqslant 0$ in $[a, b]$ we have $M_r - m_r \geqslant 0$ in every subinterval $[x_{r-1}, x_r]$ of any net, and the result follows.

This result is a particular case of the following.

Property 3′ If f and g are both integrable-R over (a, b) and if $f \geqslant g$ throughout $[a, b]$, then $\int_a^b f(x)\,dx \geqslant \int_a^b g(x)\,dx$.

Property 4 If f is integrable-R over (a, b), then so is $|f|$ and

$$\left|\int_a^b f(x)\,dx\right| \leqslant \int_a^b |f(x)|\,dx.$$

Proof. Let

$$p(x) = \begin{cases} f(x) & \text{when } f \geqslant 0, \\ 0 & \text{otherwise,} \end{cases}$$

and

$$q(x) = \begin{cases} -f(x) & \text{when } f < 0, \\ 0 & \text{otherwise.} \end{cases}$$

We can also write $p(x) = \max\{f(x), 0\}$, $q(x) = \max\{-f(x), 0\}$; then

$$f = p - q \quad \text{and} \quad |f| = p + q.$$

The proof of the integrability of p and q is left as an exercise. The inequality follows from

$$\left|\int_a^b f(x)\,dx\right| = \left|\int_a^b [p(x) - q(x)]\,dx\right|$$
$$\leqslant \left|\int_a^b p(x)\,dx\right| + \left|\int_a^b q(x)\,dx\right|,$$

using the triangular inequality.

Since both p and q are non-negative, the modulus signs may be dropped and the right-hand side written

$$\int_a^b [p(x) + q(x)]\,dx = \int_a^b |f(x)|\,dx.$$

Notice that the converse is false; a counter-example establishing this is

$$f(x) = \begin{cases} -1, & x \text{ rational,} \\ 1, & x \text{ irrational.} \end{cases}$$

Property 5 If $m \leqslant f(x) \leqslant M$ and $g \geqslant 0$ in $[a, b]$ and if g and fg are integrable-R over (a, b), then

$$m \int_a^b g(x)\,dx \leqslant \int_a^b f(x)\,g(x)\,dx \leqslant M \int_a^b g(x)\,dx.$$

roof. Clearly $(f-m)g \geqslant 0$ and $(M-f)g \geqslant 0$; the result then follows sing Property 3 above.

If f is continuous in $[a, b]$, then f takes all values between m and M. herefore, there is a point ξ for which

$$\int_a^b f(x)g(x)\,\mathrm{d}x = f(\xi)\int_a^b g(x)\,\mathrm{d}x, \qquad a \leqslant \xi \leqslant b.$$

aking $g \equiv 1$ gives a particular case of this property which is known as e *first mean value theorem for integrals*:

roperty 5′ If f is integrable-R over (a, b) and $m \leqslant f(x) \leqslant M$ in $[a, b]$, en

$$m(b-a) \leqslant \int_a^b f(x)\,\mathrm{d}x \leqslant M(b-a),$$

:

$$\int_a^b f(x)\,\mathrm{d}x = \mu(b-a),$$

r some value μ such that $m \leqslant \mu \leqslant M$.

operty 5″ If f is continuous for $[a, b]$, then $f(\xi) = \mu$ at some point ξ $[a, b]$. That is

$$\int_a^b f(x)\,\mathrm{d}x = (b-a)f(\xi), \qquad a \leqslant \xi \leqslant b.$$

operty 6 (Schwarz's inequality) If f and g are integrable-R over (a, b), en

$$\left(\int_a^b f(x)g(x)\,\mathrm{d}x\right)^2 \leqslant \int_a^b [f(x)]^2\,\mathrm{d}x \int_a^b [g(x)]^2\,\mathrm{d}x.$$

oof. $(f+\lambda g)^2 = f^2 + 2fg\lambda + g^2\lambda^2 \geqslant 0$ for all values of the constant λ. en, using Properties 2 and 3 above,

$$\lambda^2\int_a^b [g(x)]^2\,\mathrm{d}x + 2\lambda\int_a^b f(x)g(x)\,\mathrm{d}x + \int_a^b [f(x)]^2\,\mathrm{d}x \geqslant 0.$$

hwarz's inequality follows from the condition the quadratic in λ is n-negative.

ERCISES 9(c)

1. Given that f is integrable-R over (a, b) and λ is a constant, prove that

$$\int_a^b \lambda f(x)\,\mathrm{d}x = \lambda\int_a^b f(x)\,\mathrm{d}x.$$

2. Prove that, if f is integrable-R over (a, b), then it is integrable-R over (α, β), where $a \leqslant \alpha < \beta \leqslant b$.

3. Given that f is integrable-R over (a, b) and c is any point such that $a < c < b$, prove that

$$\int_a^b f(x)\,dx = \int_a^c f(x)\,dx + \int_c^b f(x)\,dx.$$

4. Given that f is integrable-R over (a, b) and $|f(x)| \geqslant m > 0$ in $[a, b]$, prove that $1/f(x)$ is integrable-R over (a, b).

5. Given that f is integrable-R over (a, b) and $|f(x)| \leqslant M$ in $[a, b]$, prove that

$$\left| \int_a^b f(x)\,dx \right| \leqslant M(b-a).$$

6. Given that f and g are bounded and integrable-R over (a, b) and $M(x) = \max \{f(x), g(x)\}$, $m(x) = \min \{f(x), g(x)\}$ for each x in $[a, b]$, prove that $M(x)$ and $m(x)$ are integrable-R over (a, b).

7. Prove that if f, f' and g are continuous in $[a, b]$ and f is monotonic, then there is a point ξ such that $a \leqslant \xi \leqslant b$ and

$$\int_a^b f(x)g(x)\,dx = f(a) \int_a^\xi g(x)\,dx + f(b) \int_\xi^b g(x)\,dx.$$

This result is known as the *second mean value theorem for integrals.*

8. Prove that if $f > 0, f' < 0$ and g are continuous in $[a, b]$ and if

$$m \leqslant \int_a^\xi g(x)\,dx \leqslant M$$

for $a \leqslant \xi \leqslant b$, then

$$mf(a) \leqslant \int_a^b f(x)g(x)\,dx \leqslant Mf(a)$$

and $\displaystyle\int_a^b f(x)g(x)\,dx = f(a) \int_a^\eta g(x)\,dx, \qquad a \leqslant \eta \leqslant b.$

This result is known as *Bonnet's form of the second mean value theorem.*

9. Verify the second mean value theorem in the cases:

(a) $\int_0^1 xe^x\,dx$ (b) $\int_0^{\pi/2} \sin x \cos x\,dx$ (c) $\int_0^1 x^4(1-x)^2\,dx$.

10. Show that, if $x > 0$, then

$$e^x - 1 < \int_0^x \sqrt{(e^{2t} + e^{-t})}\, dt < \sqrt{[(e^x - 1)(e^x - \tfrac{1}{2})]}.$$

11. Show that

$$\int_0^{2\pi} \sqrt{(1 - k^2 \sin^2\theta)}\, d\theta \leqslant 2\pi\sqrt{(1 - \tfrac{1}{2}k^2)}, \qquad 0 \leqslant k \leqslant 1.$$

12. Prove that if $0 < a < b$ and if for $a \leqslant x \leqslant b, f \geqslant 0, xf'(x) + f(x) \geqslant 0$, then

$$\left| \int_a^b f(x) \cos(\log x)\, dx \right| \leqslant 2bf(b).$$

13. Prove that if f is continuous and never negative in $[a, b]$, and $\int_a^b f(x)\, dx = 0$, then $f \equiv 0$ in $[a, b]$.

14. Substitute $t = x - u$ in the relation $f(x) = f(0) + \int_0^x f(t)\, dt$, and integrate by parts to give the Maclaurin expansion of a function f for which $f^{(n)}$ is continuous in $0 \leqslant x \leqslant h$,

$$f(x) = \sum_{r=0}^{n-1} \frac{x^r}{r!} f^{(r)}(0) + R_n,$$

where $$R_n = \frac{1}{(n-1)!} \int_0^x (x-u)^{n-1} f^{(n)}(u)\, du.$$

15. Show that the remainder in the previous question can be written

$$R_n = \frac{x^n}{(n-1)!} \int_0^1 (1-v)^{p-1}(1-v)^{n-p} f^{(n)}(xv)\, dv,$$

where p is a positive integer $\leqslant n$. Hence find the Schlömilch–Roche form of remainder,

$$R_n = \frac{x^n}{(n-1)!} \frac{(1-\theta)^{n-p}}{p} f^{(n)}(\theta x), \qquad 0 < \theta < 1.$$

Deduce the Lagrange and Cauchy forms of the remainder.

9.7 The fundamental theorem of integral calculus

In this section we develop the relation between integration and differentiation.

Theorem If f is integrable-R over (a, b) and $F(x) = \int_a^x f(t)\,\mathrm{d}t$, then F is a continuous function of x in $[a, b]$.

If, in addition, f is continuous in $[a, b]$, then F is differentiable and $F' = f$.

Proof. If x and $x + h$ are both in $[a, b]$,

$$\left|F(x+h) - F(x)\right| = \left|\int_x^{x+h} f(t)\,\mathrm{d}t\right|$$

$$\leqslant K\,|h|,$$

where $K = \sup|f|$ in $[a, b]$. Hence F is continuous in $[a, b]$.

Using the first mean value theorem for integrals,

$$\frac{F(x+h) - F(x)}{h} = f(\xi),$$

for some value ξ such that $x \leqslant \xi \leqslant x + h$. Therefore if $h \to 0+$, $\xi \to x$ and $F'_+(x) = f(x)$. Similarly if $h \to 0_-$ we have $F'_-(x) = f(x)$, and we have shown that $F'(x) = f(x)$.

This result can be written

$$\frac{\mathrm{d}}{\mathrm{d}x}\int_a^x f(t)\,\mathrm{d}t = f(x),$$

and the theorem asserts that a continuous function possesses a *primitive*. A function f possesses a primitive ψ in an interval $[a, b]$ if ψ is continuous and ψ' exists and equals f at all points of (a, b), except for at most a finite number of points. Clearly if ψ is a primitive of f, then so is $\psi + C$ for any constant C.

Example If $\psi(x) = x^2 \sin(1/x^2), x \neq 0, \psi(0) = 0$, then $\psi'(0) = 0$ and for $x \neq 0$,

$$\psi'(x) = 2x \sin(1/x^2) - 2x^{-1}\cos(1/x^2).$$

That is, $f \equiv \psi'$ has, in any finite interval, a primitive given by ψ. Notice, however, that f is not integrable in any interval containing the origin, as it is unbounded in every neighbourhood of the origin.

Theorem If f is continuous in $[a, b]$ and ψ is a primitive of f, then

$$\int_a^x f(x)\,\mathrm{d}x = \psi(b) - \psi(a).$$

Proof. Let $F(x) = \int_a^x f(t)\,\mathrm{d}t$ and $g(x) = F(x) - \psi(x)$. Then $g'(x) \equiv 0$ in $[a, b]$ and so g is a constant, say $g(a)$. That is, $F(x) = \psi(x) + g(a)$, 'here

$$g(a) = F(a) - \psi(a) = -\psi(a),$$

nce $F(a) = 0$. Hence $F(x) = \psi(x) - \psi(a)$ and the result follows on etting $x = b$.

XERCISES 9(d)

1. Prove that if f is integrable-R over (a, b) and has a primitive ψ, then

$$\int_a^b f(x)\,dx = \psi(b) - \psi(a).$$

2. Given that f is continuous in $[a, b]$ and f' exists in (a, b) but not necessarily at $x = a$ or $x = b$, prove that, if f' is integrable-R over (a, b), then

$$\int_a^b f'(x)\,dx = f(b) - f(a).$$

3. Given that $f(x) = x \sin(\log x)$, $0 < x \leqslant 1$ and $f(0) = 0$, show that f is continuous in $[0, 1]$ and f' exists in $(0, 1)$. Show also that $\int_0^1 f'(x)\,dx = 0$.

4. Show that $\psi(x) = x^{3/2} \sin(1/x)$ has an unbounded derivative in $[0, 1]$. Deduce that $\int_a^b \psi'(x)\,dx$ does not exist if the origin is in $[a, b]$.

.8 Infinite integrals

the definition of the definite integral $\int_a^b f(x)\,dx$ given in the previous ctions, we have assumed that the interval (a, b) is finite and that f is ounded throughout the interval. We now examine the effect of relaxing ese conditions.

First we consider an infinite interval, Let f be bounded and tegrable-R over (a, λ) for every $\lambda > a$; then we define the *infinite tegral*

$$\int_a^\infty f(x)\,dx = \lim_{\lambda\to\infty} \int_a^\lambda f(x)\,dx,$$

hen this limit exists. If the limiting value is L we say the integral *nverges to L*. If the limit does not exist the integral is said to *diverge.* sing the definition of a limit we can write:

$$\text{if } \forall\epsilon > 0;\ \exists X.\ \left| L - \int_a^\lambda f(x)\,dx \right| < \epsilon \qquad \text{for } \lambda > X,$$

en $\int_a^\infty f(x)\,dx$ converges to L.

We can also express this result in a way analogous to the general principle of convergence of series.

A necessary and sufficient condition for the convergence of $\int_a^\infty f(x)\,dx$ is

$$\forall \epsilon > 0;\ \exists X.\ \left|\int_{\lambda'}^{\lambda''} f(x)\,dx\right| < \epsilon \qquad \text{for } \lambda'' > \lambda' > X.$$

If f is bounded and integrable-R over $(-\lambda, a)$ for every $\lambda > a$, then we define

$$\int_{-\infty}^{a} f(x)\,dx = \lim_{\lambda\to\infty} \int_{-\lambda}^{a} f(x)\,dx,$$

when this limit exists. Also

$$\int_{-\infty}^{\infty} f(x)\,dx = \int_{-\infty}^{a} f(x)\,dx + \int_{a}^{\infty} f(x)\,dx$$

provided *both* integrals on the right-hand side of the relation exist.

Example 1 $\displaystyle\int_0^\infty \frac{dx}{1+x^2}$ converges to $\frac{1}{2}\pi$.

For $$\int_0^\lambda \frac{dx}{1+x^2} = \tan^{-1}\lambda \to \tfrac{1}{2}\pi \qquad \text{as } \lambda\to\infty.$$

Example 2 $\int_{-\infty}^{0} e^x\,dx$ converges to 1.

For $$\int_{-\lambda}^{0} e^x\,dx = 1 - e^{-\lambda} \to 1 \qquad \text{as } \lambda\to\infty.$$

Example 3 $\int_1^\infty x^{-\alpha}\,dx$ converges if $\alpha > 1$ and diverges if $\alpha \leqslant 1$.

For $$\int_1^\lambda x^{-\alpha}\,dx = (\lambda^{1-\alpha}-1)/(1-\alpha), \qquad \alpha \neq 1$$

$$\to \begin{cases} 1/(\alpha-1) & \text{as } \lambda\to\infty \text{ if } \alpha > 1; \\ \infty & \text{as } \lambda\to\infty \text{ if } \alpha < 1. \end{cases}$$

When $\alpha = 1$,
$$\int_1^\lambda x^{-1}\,dx = \log\lambda \to \infty \qquad \text{as } \lambda\to\infty.$$

Example 4 $\int_0^\infty (\sin x)/x\,dx$ is convergent.

Using the second mean value theorem for integrals (Exercises 9(c), No. 7

$$\int_{\lambda'}^{\lambda''} \frac{\sin x\,dx}{x} = \frac{1}{\lambda'}\int_{\lambda'}^{\xi} \sin x\,dx + \frac{1}{\lambda''}\int_{\xi}^{\lambda''} \sin x\,dx,$$

$$0 < \lambda' \leqslant \xi \leqslant \lambda''.$$

Also

$$\left|\int_a^b \sin x \, dx\right| = \left|\cos a - \cos b\right| \leqslant \left|\cos a\right| + \left|\cos b\right| \leqslant 2,$$

for any interval (a, b). Then

$$\left|\int_{\lambda'}^{\lambda''} \frac{\sin x}{x} \, dx\right| \leqslant 2\left|\frac{1}{\lambda'} + \frac{1}{\lambda''}\right|$$

$$\leqslant \frac{4}{\lambda''}, \quad \lambda'' \geqslant \lambda',$$

$$< \epsilon \qquad \text{for } \lambda'' \geqslant \lambda' > X = 4/\epsilon.$$

Therefore $\int_0^\infty (\sin x)/x \, dx$ converges.

Example 5

$$\int_{-\infty}^{\infty} \frac{dx}{1+x^2} = \int_{-\infty}^{a} \frac{dx}{1+x^2} + \int_a^\infty \frac{dx}{1+x^2}$$

$$= \lim_{\lambda \to \infty} \int_{-\lambda}^{a} \frac{dx}{1+x^2} + \lim_{\mu \to \infty} \int_a^\mu \frac{dx}{1+x^2}$$

$$= \lim_{\lambda \to \infty} (\tan^{-1} a + \tan^{-1} \lambda) + \lim_{\mu \to \infty} (\tan^{-1} \mu - \tan^{-1} a)$$

$$= \pi.$$

The existence of $\lim_{\lambda \to \infty} \int_{-\lambda}^{\lambda} f(x) \, dx$ does not imply that $\int_{-\infty}^{\infty} f(x) \, dx$ converges. Thus $\int_{-\infty}^{\infty} \sin x \, dx$ does not exist since neither $\int_{-\infty}^{a} \sin x \, dx$ nor $\int_a^\infty \sin x \, dx$ exists. But $\int_{-\lambda}^{\lambda} \sin x \, dx = 0$ and so $\lim_{\lambda \to \infty} \int_{-\lambda}^{\lambda} \sin x \, dx = 0$. In such cases $\lim_{\lambda \to \infty} \int_{-\lambda}^{\lambda} f(x) \, dx$ is called the Cauchy principal value of the infinite integral and is written

$$P\int_{-\infty}^{\infty} f(x) \, dx.$$

Comparison test for integrals If f and g are positive, bounded functions integrable-R over any finite interval and $0 < f(x) < g(x)$ for $x \geqslant a$, then

$$\int_a^\infty f(x) \, dx \text{ converges if } \int_a^\infty g(x) \, dx \text{ converges,}$$

and

$$\int_a^\infty g(x) \, dx \text{ diverges if } \int_a^\infty f(x) \, dx \text{ diverges.}$$

Proof. $\int_a^\infty g(x)\,dx = L$; then

$$\int_a^\lambda f(x)\,dx \leqslant \int_a^\lambda g(x)\,dx \leqslant L \qquad \text{for } \lambda \geqslant a,$$

since $f(x) < g(x)$ for $x \geqslant a$. Therefore $\int_a^\infty f(x)\,dx$ converges (see Exercises 9(e), No. 2). Also, since $\int_a^\lambda g(x)\,dx \geqslant \int_a^\lambda f(x)\,dx$ for $\lambda > a$, then $\int_a^\infty g(x)\,dx$ diverges when $\int_a^\infty f(x)\,dx$ is divergent.

A useful integral for comparison purposes is that given in Example 3 above – for any $a > 0$, $\int_a^\infty x^{-\alpha}\,dx$ is convergent for $\alpha > 1$ and divergent for $\alpha \leqslant 1$.

Example 6 $\displaystyle\int_1^\infty \frac{dx}{x\sqrt{(1+x^2)}}$

This converges by comparison with the convergent integral $\int_1^\infty x^{-2}\,dx$, since

$$\frac{1}{x\sqrt{(1+x^2)}} < \frac{1}{x^2} \qquad \text{for } x > 1.$$

Example 7 $\displaystyle\int_2^\infty \frac{dx}{\sqrt{(x^2-1)}}$ diverges.

For

$$\frac{1}{\sqrt{(x^2-1)}} > \frac{1}{x} \qquad (x \geqslant 2)$$

and $\int_2^\infty (1/x^2)\,dx$ diverges.

If f is bounded and integrable-R over (a, λ) for any $\lambda > a$, and $\int_a^\infty |f(x)|\,dx$ converges, then $\int_a^\infty f(x)\,dx$ is said to be *absolutely convergent*. Using Property 4, Section 9.6, we have

$$\left| \int_{\lambda'}^{\lambda''} f(x)\,dx \right| \leqslant \int_{\lambda'}^{\lambda''} \left| f(x) \right| dx,$$

and so *an absolutely convergent integral is convergent.* Example 9 below shows that the converse is false.

Example 8 $\displaystyle\int_0^\infty \frac{\cos x}{1+x^2}\,dx$ converges absolutely.

For

$$\left| \int_0^\lambda \frac{\cos x}{1+x^2}\,dx \right| \leqslant \int_0^\lambda \frac{|\cos x|}{1+x^2}\,dx \leqslant \int_0^\lambda \frac{dx}{1+x^2},$$

and the result follows by comparison.

Example 9 $\displaystyle\int_0^\infty \frac{\sin x}{x}\,dx$ is convergent but not absolutely convergent.

The convergence was proved in Example 4 above.

$$\int_0^\lambda \frac{|\sin x|}{x}\,dx > \int_0^{n\pi} \frac{|\sin x|}{x}\,dx, \qquad \lambda > n\pi.$$

Now

$$\int_0^{n\pi} \frac{|\sin x|}{x}\,dx = \sum_{r=1}^{n} \int_{(r-1)\pi}^{r\pi} \frac{|\sin x|}{x}\,dx = \sum_{r=1}^{n} \int_0^{\pi} \frac{\sin y}{(r-1)\pi + y}\,dy,$$

putting $x = y + (r-1)\pi$. Also

$$\frac{1}{(r-1)\pi + y} \geqslant \frac{1}{r\pi}, \qquad 0 \leqslant y \leqslant \pi.$$

Then $$\int_0^\lambda \frac{|\sin x|}{x}\,dx > \sum_{r=1}^{n} \frac{1}{r\pi} \int_0^{\pi} \sin y\,dy = \frac{2}{\pi} \sum_{r=1}^{n} \frac{1}{r}.$$

Therefore, since $\Sigma_{r=1}^{n} (1/r)$ is unbounded, $\int_0^\infty |\sin x|/x\,dx$ diverges.

9.9 Improper integrals

In the previous section we considered integrals of bounded functions over an infinite interval, in this section we take a finite interval but allow the function to become unbounded in the interval. Some writers call such integrals *improper* to distinguish them from the infinite-interval types.

Suppose f has an infinite discontinuity at $x = a$ and is bounded and integrable over $(a + \xi, b)$, where $a < a + \xi < b$; then if $\lim_{\xi \to 0} \int_{a+\xi}^{b} f(x)\,dx$ exists we say that $\int_a^b f(x)\,dx$ converges. Similarly, if f has an infinite discontinuity at $x = b$ and is bounded and integrable over $(a, b - \xi)$, where $a < b - \xi < b$, then $\int_a^b f(x)\,dx$ converges if $\lim_{\xi \to 0} \int_a^{b-\xi} f(x)\,dx$ exists.

These results can be expressed in terms of ϵ. Thus $\int_a^b f(x)\,dx$ converges to L if

$$\forall \epsilon > 0;\ \exists \delta\ .\left| L - \int_{a+\xi}^{b} f(x)\,dx \right| < \epsilon \qquad \text{for } 0 < \xi \leqslant \delta.$$

Further, $\int_a^b f(x)\,dx$ is convergent if

$$\forall \epsilon > 0;\ \exists \delta\ .\left| \int_{a+\xi'}^{a+\xi''} f(x)\,dx \right| < \epsilon \qquad \text{for } 0 < \xi' < \xi'' \leqslant \delta.$$

If f is bounded in $[a, b]$ *except* at the one point $c, a < c < b$, then if f is integrable over any sub-interval not containing c, the two integrals

$$\int_a^{c-\xi'} f(x)\,\mathrm{d}x \quad \text{and} \quad \int_{c+\xi''}^{b} f(x)\,\mathrm{d}x$$

both exist for $a < c - \xi' < c, c < c + \xi'' < b$. If *both* integrals converge as $\xi', \xi'' \to 0$ *independently* we define

$$\int_a^b f(x)\,\mathrm{d}x = \lim_{\xi' \to 0} \int_a^{c-\xi'} f(x)\,\mathrm{d}x + \lim_{\xi'' \to 0} \int_{c+\xi''}^{b} f(x)\,\mathrm{d}x.$$

This result can be extended to a finite number of points of infinite discontinuities in the interval (a, b).

Example 1 $\int_0^1 \frac{\mathrm{d}x}{\sqrt{(1-x^2)}}$ converges.

For $$\int_0^{1-\xi} \frac{\mathrm{d}x}{\sqrt{(1-x^2)}} = \sin^{-1}(1-\xi) \to \tfrac{1}{2}\pi \text{ as } \xi \to 0.$$

Example 2 $\int_0^1 x^{-\alpha}\,\mathrm{d}x$ converges if $\alpha < 1$ and diverges if $\alpha \geqslant 1$.

For $\int_\xi^1 x^{-\alpha}\,\mathrm{d}x = (1-\xi^{1-\alpha})/(1-\alpha), \qquad \alpha \neq 1$

$$\to \begin{cases} 1/(1-\alpha) & \text{as } \xi \to 0 \text{ if } \alpha < 1, \\ \infty & \text{as } \xi \to 0 \text{ if } \alpha > 1. \end{cases}$$

When $\alpha = 1$,

$$\int_\xi^1 x^{-1}\,\mathrm{d}x = -\log \xi \to \infty \qquad \text{as } \xi \to 0.$$

If the only infinite discontinuity of f is at $x = a$, $\int_a^b f(x)\,\mathrm{d}x$ is said to be *absolutely convergent* if f is bounded and integrable over $(a+\xi, b)$, where $a < a + \xi < b$, and $\int_a^b |f(x)|\,\mathrm{d}x$ converges. A comparison test analogous to that in the previous section can be formulated and the integral $\int_a^b (x-a)^{-\alpha}\,\mathrm{d}x$ can be used for comparison. Thus we have the following.

α-test for integrals If f is bounded and integrable over the arbitrary interval $(a + \xi, b)$, and if

$$\lim_{x \to a+} (x-a)^{\alpha} f(x) = l,$$

then $\int_a^b f(x)\,\mathrm{d}x$ is absolutely convergent if $0 < \alpha < 1$; and if $l \neq 0$, the integral is divergent if $\alpha \geqslant 1$.

Proof. If $\lim_{x \to a+} (x-a)^\alpha f(x) = l$, then

$$\tfrac{1}{2}|l| < (x-a)^\alpha |f(x)| < \tfrac{3}{2}|l|, \qquad \text{for } a < x < a+\delta$$

(see Section 2.6). Thus

$$\tfrac{1}{2}|l| \int_{a+\xi'}^{a+\xi''} (x-a)^{-\alpha}\,\mathrm{d}x < \int_{a+\xi'}^{a+\xi''} |f(x)|\,\mathrm{d}x < \tfrac{3}{2}|l| \int_{a+\xi'}^{a+\xi''} (x-a)^{-\alpha}\,\mathrm{d}x,$$
$$0 < \xi' < \xi'' \leqslant \delta,$$

and the result follows.

Example 3

$$\int_0^1 \frac{\mathrm{d}x}{\sqrt{[x(1-x)]}} = \int_0^a \frac{\mathrm{d}x}{\sqrt{[x(1-x)]}} + \int_a^1 \frac{\mathrm{d}x}{\sqrt{[x(1-x)]}}.$$

Here $f(x) = [x(1-x)]^{-1/2}$ and $x^{1/2}f(x) \to 1$ as $x \to 0$. Therefore

$$\int_0^a \frac{\mathrm{d}x}{\sqrt{[x(1-x)]}}$$

converges; similarly

$$\int_a^1 \frac{\mathrm{d}x}{\sqrt{[x(1-x)]}}$$

converges.

Example 4 $\Gamma(x) = \int_0^\infty e^{-t}t^{x-1}\,\mathrm{d}t$.

Since

$$\lim_{t \to \infty} t^2(e^{-t}t^{x-1}) = 0,$$

we have

$$0 < e^{-t}t^{x-1} < At^{-2}, \qquad t \geqslant 1,$$

and so $\int_1^\infty e^{-t}t^{x-1}\,\mathrm{d}t$ converges by comparison with $\int_1^\infty t^{-2}\,\mathrm{d}t$ for all values of x. When $x \geqslant 1$ the integrand is bounded at $t = 0$; therefore $\int_0^\infty e^{-t}t^{x-1}\,\mathrm{d}t$ converges for $x \geqslant 1$.

Now consider $0 < x < 1$; the integrand has an infinity at the origin $t = 0$. Choose α such that $0 < \alpha < 1$ and $\alpha + x > 1$; then

$$t^\alpha(e^{-t}t^{x-1}) \to 0 \qquad \text{as } t \to 0,$$

and $\int_0^1 e^{-t}t^{x-1}\,\mathrm{d}t$ converges by comparison with $\int_0^1 t^{-\alpha}\,\mathrm{d}t$ for $0 < \alpha < 1$. It was shown above that $\int_1^\infty e^{-t}t^{x-1}\,\mathrm{d}t$ converges, therefore $\int_0^\infty e^{-t}t^{x-1}\,\mathrm{d}t$ converges when $0 < x < 1$.

When $x \leqslant 0$, the integrand has an infinite discontinuity at $t = 0$. Since

$$\lim_{t \to 0} t^{1-x}(e^{-t}t^{x-1}) = 1$$

and $\int_0^1 t^{x-1}\, dt$ diverges when $x \leqslant 0$, then $\int_0^1 e^{-t}t^{x-1}\, dt$ diverges by comparison.

We have shown that the *gamma function* $\Gamma(x) = \int_0^\infty e^{-t}t^{x-1}\, dt$ converge for $x > 0$.

EXERCISES 9(e)

1. Decide whether the following integrals converge or diverge; when the integral converges evaluate it:

(a) $\int_0^\infty e^{-2x}\, dx$ (b) $\int_1^\infty x^{-2/3}\, dx$ (c) $\int_0^1 x^{-2/3}\, dx$

(d) $\displaystyle\int_2^\infty \frac{dx}{1-x}$ (e) $\displaystyle\int_1^2 \frac{dx}{1-x}$ (f) $\displaystyle\int_0^{\sqrt{2}} \frac{dx}{\sqrt{(2-x^2)}}$

(g) $\int_0^\infty \sin x\, dx$ (h) $\int_{-\infty}^0 e^{ax}\, dx$ (i) $\int_{-\infty}^\infty \operatorname{sech} x\, dx$

(j) $\displaystyle\int_{-\infty}^\infty \frac{dx}{1+3\operatorname{ch} x}$ (k) $\int_0^1 x^{-1/2}(1-x)^{-1/2}\, dx$.

2. Show that if $f > 0$ for $x \geqslant a$, then

$$F(x) = \int_a^x f(t)\, dt$$

is a monotonic increasing function of x, and that if F is bounded then $\int_a^\infty f(x)\, dx$ converges.

3. By writing $p = |f| + f$ and $\alpha = |f| - f$, prove that an absolutely convergent integral converges.

4. Prove that if f is an odd function of x, integrable over every finite interval, then $P\int_{-\infty}^\infty f(x)\, dx = 0$.

5. Prove that, if f is bounded and integrable over any finite interval (a, b), then if $\lim_{x\to\infty} x^\alpha f(x) = l$, $\int_a^\infty f(x)\, dx$, $(a > 0)$ converges if $\alpha > 1$; if $\alpha \leqslant 1$ and $l \neq 0$, $\int_a^\infty f(x)\, dx$ diverges. (*The α-test for integrals.*)

6. Prove that if f is bounded and monotonic for $x \geqslant a$ and g is bounded and integrable over any finite interval (a, b), and $\int_a^\infty g(x)\, dx$ converge then $\int_a^\infty f(x)g(x)\, dx$ converges.

7. Test the following integrals for convergence/divergence:

(a) $\displaystyle\int_1^\infty \frac{dx}{1+x^3}$ (b) $\displaystyle\int_{-\infty}^0 \frac{\sin x}{1+x^2}\, dx$ (c) $\displaystyle\int_1^\infty \frac{\sin^2 x}{x^3}\, dx$

(d) $\int_0^1 e^{1/x}\, dx$ (e) $\int_0^\infty e^{-x} \sin x \frac{dx}{x}$ (f) $\int_0^\infty \frac{x^2\, dx}{(a^2+x^2)^2}$

(g) $\int_0^\infty \frac{x^3\, dx}{(a^2+x^2)^2}$ (h) $\int_0^1 \frac{dx}{x(x+1)}$ (i) $\int_0^{\pi/2} \frac{dx}{\sqrt{(\sin x)}}$

(j) $\int_0^1 x^{\alpha-1}(1-x)^{\beta-1}\, dx$ (k) $\int_0^1 \sin(1/x) \frac{dx}{x}$

(l) $\int_0^\infty \frac{\sin x}{x^\alpha}\, dx$ (m) $\int_0^\infty \sin(x^2)\, dx$.

8. The function f whose graph is composed of triangular pulses of unit height at $x = n$ is shown in Fig. 7, the base of the triangle with peak at $x = n$ being $2/(n+1)^2$. Show that $\int_0^\infty f(x)\, dx$ converges, although $f(x) \nrightarrow 0$ as $x \to \infty$.

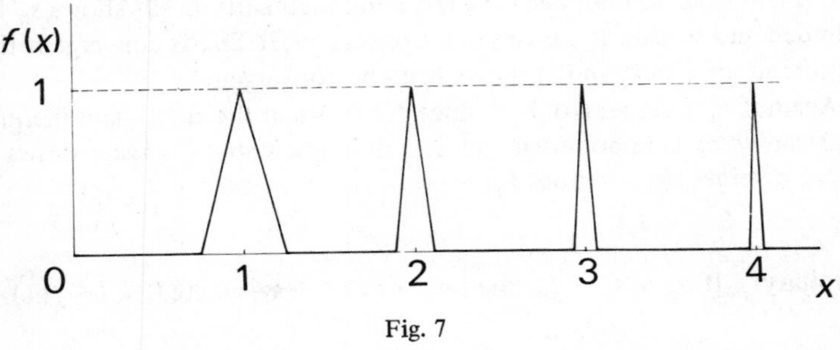

Fig. 7

9. Construct an example to show that $\int_0^\infty f(x)\, dx$ may converge even when $f(x)$ is unbounded as $x \to \infty$.

10. Prove that if f is a positive, decreasing function for $x > a$ and $f(x) \to 0$ as $x \to \infty$, then

$$\int_a^\infty f(x) \cos \lambda x\, dx$$

is absolutely convergent for $\lambda > 0$.

9.10 The integral test

Theorem If f is a positive, monotonic decreasing function of x in $(1, \infty)$, then $\Sigma f(n)$ and $\int_1^\infty f(x)\, dx$ converge or diverge together.

That is, if $s_n = \Sigma_{r=1}^n f(r)$ and $I_n = \int_1^n f(x)\, dx$, then the sequences (s_n) and (I_n) either both converge or both diverge.

Proof. Let $r-1<x\leqslant r$; then

$$f(r)\leqslant f(x)\leqslant f(r-1).$$

Integrating these inequalities over the interval $[r-1, r]$, we have

$$f(r)\leqslant \int_{r-1}^{r} f(x)\,\mathrm{d}x\leqslant f(r-1). \tag{1}$$

Then adding these inequalities for $r=2, 3, \ldots, n$ gives

$$\sum_{r=2}^{n} f(r)\leqslant \int_{1}^{n} f(x)\,\mathrm{d}x\leqslant \sum_{r=1}^{n-1} f(r),$$

or

$$\begin{aligned} s_n - f(1) \leqslant I_n &\leqslant s_{n-1} \\ &\leqslant s_n, \qquad \text{since } f(n)>0. \end{aligned} \tag{2}$$

Since $f>0$, both s_n and I_n are monotonic increasing. Then, if I_n is convergent (and so bounded) the left-hand inequality in (2) shows s_n is bounded and so Σu_n is convergent. Conversely, if Σu_n is convergent the right-hand inequality in (2) shows I_n to be convergent.

Again if I_n diverges (to $+\infty$, since $f>0$), then the right-hand inequality in (2) shows s_n is unbounded and Σu_n diverges. Using (2) again shows that if Σu_n diverges, so also does I_n.

Corollary If $\sigma_n = s_n - I_n$, then $\sigma_n \to l$ as $n\to\infty$, where $0\leqslant l\leqslant f(1)$.

Proof. σ_n is monotonic decreasing and bounded, since (2) can be re-written $0\leqslant \sigma_n \leqslant f(1)$, and using (1), we have

$$\sigma_n - \sigma_{n+1} = \int_{n}^{n+1} f(x)\,\mathrm{d}x - f(n+1)\geqslant 0.$$

In particular, when $f(x)=1/x$,

$$\sigma_n = 1+\frac{1}{2}+\frac{1}{3}+\cdots+\frac{1}{n}-\log n\to\gamma \qquad \text{as } n\to\infty,$$

$\gamma = 0{\cdot}5772\ldots$ is known as *Euler's constant.*

The inequalities (1) above also provide a way of estimating bounds for partial sums of some series. For, summing inequalities (1) for $r=n+1$, $n+2, \ldots, n+p$ gives

$$\sum_{r=n+1}^{n+p} f(r)\leqslant \int_{n}^{n+p} f(x)\,\mathrm{d}x\leqslant \sum_{r=n}^{n+p-1} f(r);$$

hence $$\int_{n+1}^{n+p+1} f(x)\,\mathrm{d}x \leqslant \sum_{r=n+1}^{n+p} f(r) \leqslant \int_{n}^{n+p} f(x)\,\mathrm{d}x. \tag{3}$$

Also from (2) $$I_n \leqslant s_n \leqslant I_n + f(1). \tag{4}$$

Example 1 $\Sigma n^{-\alpha}$ is convergent for $\alpha > 1$ and divergent for $\alpha \leqslant 1$.

Let $f(x) = x^{-\alpha}$; then

$$\int_1^n x^{-\alpha}\,\mathrm{d}x = \begin{cases} (n^{1-\alpha} - 1)/(1-\alpha), & \alpha \neq 1, \\ \log n, & \alpha = 1, \end{cases}$$

and the result follows.

Example 2 Let $f(x) = 1/x^2$; then from (3)

$$\int_{n+1}^{n+p+1} \frac{\mathrm{d}x}{x^2} \leqslant \sum_{r=n+1}^{n+p} \frac{1}{r^2} \leqslant \int_n^{n+p} \frac{\mathrm{d}x}{x^2}.$$

That is

$$\frac{p}{(n+1)(n+p+1)} \leqslant \frac{1}{(n+1)^2} + \frac{1}{(n+2)^2} + \cdots + \frac{1}{(n+p)^2} \leqslant \frac{p}{n(n+p)}.$$

As the series Σn^{-2} converges we can fix n and let $p \to \infty$ to give

$$\frac{1}{n+1} \leqslant \frac{1}{(n+1)^2} + \frac{1}{(n+2)^2} + \frac{1}{(n+3)^2} + \cdots \leqslant \frac{1}{n}.$$

Alternatively, using (4)

$$\int_1^n \frac{\mathrm{d}x}{x^2} \leqslant \sum_{r=1}^{n} \frac{1}{r^2} \leqslant \int_1^n \frac{\mathrm{d}x}{x^2} + 1,$$

and letting $n \to \infty$, we have

$$1 \leqslant \sum_1^\infty \frac{1}{n^2} \leqslant 2.$$

Example 3 To show that $\lim\limits_{n\to\infty} \sum\limits_{r=1}^{n} \dfrac{1}{n+r} = \log 2.$

We can write the sum as the difference between two partial sums of the

$$\gamma = \lim_{n\to\infty} (1 + \tfrac{1}{2} + \tfrac{1}{3} + \cdots + \tfrac{1}{n} - \log n).$$

That is

$$1 + \tfrac{1}{2} + \tfrac{1}{3} + \cdots + \tfrac{1}{n} = \gamma + \log n + \epsilon_n,$$

where $\epsilon_n \to 0$ as $n \to \infty$. Now

$$\sum_{r=1}^{n} \frac{1}{n+r} = \sum_{r=1}^{2n} \frac{1}{r} - \sum_{r=1}^{n} \frac{1}{r}$$
$$= \gamma + \log 2n + \epsilon_{2n} - \gamma - \log n - \epsilon_n,$$

where $\epsilon_n \to 0$ and $\epsilon_{2n} \to 0$ as $n \to \infty$. That is

$$\sum_{r=1}^{n} \frac{1}{n+r} = \log 2 + \epsilon_{2n} - \epsilon_n$$
$$\to \log 2 \qquad \text{as } n \to \infty.$$

Example 4 $1 - \frac{1}{2} - \frac{1}{4} + \frac{1}{3} - \frac{1}{6} - \frac{1}{8} + \frac{1}{5} - \cdots$ is a rearrangement of the series $1 - \frac{1}{2} + \frac{1}{3} - \frac{1}{4} + \cdots$ and its sum can be found by using Euler's constant.

Since each positive term is followed by two negative terms we consider the sum of the first $3n$ terms.

$$s_{3n} = 1 - \frac{1}{2} - \frac{1}{4} + \frac{1}{3} - \frac{1}{6} - \frac{1}{8} + \cdots + \frac{1}{2n-1} - \frac{1}{4n-2} - \frac{1}{4n}$$
$$= \left(1 + \frac{1}{3} + \frac{1}{5} + \cdots + \frac{1}{2n-1}\right) - \left(\frac{1}{2} + \frac{1}{4} + \frac{1}{6} + \cdots + \frac{1}{4n}\right)$$
$$= \left(1 + \frac{1}{2} + \frac{1}{3} + \frac{1}{4} + \cdots + \frac{1}{2n-1} + \frac{1}{2n}\right) - \left(\frac{1}{2} + \frac{1}{4} + \frac{1}{6} + \cdots + \right.$$
$$\left. - \frac{1}{2}\left(1 + \frac{1}{2} + \frac{1}{3} + \cdots + \frac{1}{2n}\right)\right)$$
$$= \frac{1}{2}\left(1 + \frac{1}{2} + \frac{1}{3} + \cdots + \frac{1}{2n}\right) - \frac{1}{2}\left(1 + \frac{1}{2} + \frac{1}{3} + \cdots + \frac{1}{n}\right)$$
$$= \tfrac{1}{2}(\gamma + \log 2n + \epsilon_{2n}) - \tfrac{1}{2}(\gamma + \log n + \epsilon_n),$$

where $\epsilon_n, \epsilon_{2n} \to 0$ as $n \to \infty$.

That is

$$s_{3n} = \tfrac{1}{2}\log 2 + \tfrac{1}{2}(\epsilon_{2n} - \epsilon_n)$$
$$\to \tfrac{1}{2}\log 2 \qquad \text{as } n \to \infty.$$

Now $s_{3n+1} = s_{3n} + \dfrac{1}{2n+1}$ and $s_{3n+2} = s_{3n} + \dfrac{1}{2n+1} - \dfrac{1}{4n+2}$;

hence $$\lim_{n\to\infty} s_{3n+1} = \lim_{n\to\infty} s_{3n} = \lim_{n\to\infty} s_{3n+2}.$$

Therefore $1-\frac{1}{2}-\frac{1}{4}+\frac{1}{3}-\cdots$ converges to $\frac{1}{2}\log 2$.

EXERCISES 9(f)

1. Test the convergence or divergence of the series Σu_n when u_n is given by:

(a) $\dfrac{1}{n^\alpha}$ (b) $\dfrac{1}{n^2+1}$ (c) $\dfrac{n}{n^2+1}$

(d) $\dfrac{1}{n(n+1)}$ (e) $\dfrac{1}{n(\log n)^\alpha}$ (f) $\dfrac{1}{1+\sqrt{n}}$.

2. Show that

(a) $2(\sqrt{n}-1)\leqslant 1+\dfrac{1}{\sqrt{2}}+\dfrac{1}{\sqrt{3}}+\cdots+\dfrac{1}{\sqrt{n}}\leqslant 2\sqrt{n}-1,$

(b) $\frac{1}{2}\pi\leqslant\dfrac{1}{2\sqrt{1}}+\dfrac{1}{3\sqrt{2}}+\dfrac{1}{4\sqrt{3}}+\cdots\leqslant\frac{1}{2}(\pi+1).$

3. Show that

$$\sum_{n=1}^{\infty}\frac{1}{n^2+1}\leqslant\tfrac{1}{2}+\tfrac{1}{4}\pi.$$

4. Find the limits as $n\to\infty$ of

(a) $\dfrac{1}{2n+1}+\dfrac{1}{2n+2}+\dfrac{1}{2n+3}+\cdots+\dfrac{1}{3n},$

(b) $1-\dfrac{1}{2}+\dfrac{1}{3}-\dfrac{1}{4}+\cdots-\dfrac{1}{2n},$

(c) $1+\dfrac{1}{3}-\dfrac{1}{2}+\dfrac{1}{5}+\dfrac{1}{7}-\dfrac{1}{4}+\cdots+\dfrac{1}{4n-3}+\dfrac{1}{4n-1}-\dfrac{1}{2n}.$

5. Show that

(a) $1+\frac{1}{3}+\frac{1}{5}-\frac{1}{2}-\frac{1}{4}+\cdots=\frac{1}{2}\log 6,$

(b) $1-\frac{1}{2}-\frac{1}{4}-\frac{1}{6}-\frac{1}{8}+\cdots=0,$

(c) $1+\frac{1}{3}+\frac{1}{5}+\frac{1}{7}+\frac{1}{9}-\frac{1}{2}-\frac{1}{4}-\frac{1}{6}+\cdots=\frac{1}{2}\log(20/3),$

the series all being rearrangements of the series $1-\frac{1}{2}+\frac{1}{3}-\frac{1}{4}+\cdots$.

6. Show that

$$\int_1^n \log x\,dx<\sum_{r=2}^{n}\log r<\int_1^n \log x\,dx+\log n.$$

Hence show that $\lim_{n\to\infty}(n!)^{1/n}/n=e^{-1}$.

CHAPTER 10

Functions of Several Variables

10.1 Functions

In this chapter we consider functions of more than one real variable. Most of the definitions and theorems apply to functions of n independent variables, but for the sake of simplicity we shall, in the main, restrict ourselves to functions of two independent variables.

Let X and Y be two sets of real numbers; then the set E of ordered pairs (x, y) with $x \in X, y \in Y$ is the *product set* $X \times Y$; that is E is a set of points P in the (x, y) plane. A function f whose domain is E is called a *function of two variables*. The value of f at P is denoted by $f(x, y) = z$ (say). The function is, therefore, the ordered pair $((x, y), z)$, usually written as an ordered triple (x, y, z). The values of the function $f(x, y)$ may be represented graphically by points on the surface $z = f(x, y)$. It is convenient to denote both the function and its value by the same letter, for example $z = z(x, y)$.

10.2 Limits

The function f tends to a limit l as (x, y) approaches (a, b) if the difference between l and the values taken by the function can be made arbitrarily small for all points sufficiently close to (a, b). That is

$$\forall \epsilon > 0; \exists \delta . \; |f(x, y) - l| < \epsilon \qquad \text{for all } (x, y) \text{ such that}$$
$$0 < \sqrt{[(x-a)^2 + (y-b)^2]} < \delta .$$

The circular neighbourhood $(x-a)^2 + (y-b)^2 < \delta^2$ can be replaced by a square neighbourhood $|x - a| + |y - b| < \delta$.

We write

$$\lim_{(x,y)\to(a,b)} f(x,y) = l \quad \text{or} \quad \lim_{\substack{x\to a\\ y\to b}} f(x,y) = l,$$

and say that l is the *double limit* of f as $(x,y)\to(a,b)$.

Example 1 $f(x,y)=(x+y)\sin(x+y)\to 0$ as $(x,y)\to(0,0)$.

$$\begin{aligned}|f(x,y)-l| &= |x+y|\,|\sin(x+y)|\\ &\leqslant |x+y|\\ &\leqslant |x|+|y|<\epsilon \qquad \text{for } |x|,|y|<\delta=\tfrac{1}{2}\epsilon.\end{aligned}$$

Example 2 $f(x,y)=\dfrac{xy}{x^2+y^2}$, $\quad x^2+y^2\neq 0$.

In this case the limiting value depends on the path along which the origin is approached. Consider an approach along the straight line $y=mx$. Then $f(x,mx)=m/(1+m^2)$, and this function of m takes all values between $\pm\frac{1}{2}$ for $-\infty<m<\infty$. Therefore

$$\lim_{(x,y)\to(a,b)} f(x,y)$$

does not exist.

Example 3 $f(x,y)=\dfrac{xy^2}{x^2+y^4}$, $\quad (x,y)\neq(0,0)$.

In this case $f\to 0$ as the origin is approached along any straight line $y=mx$, for

$$f(x,mx) = \frac{mx}{1+m^4x^2}\to 0 \qquad \text{as } x\to 0.$$

But the double limit does not exist, because the limiting value is not the same along *every* path approaching the origin. For example, along $y^2=x$, $f(y^2,y)=\frac{1}{2}$.

10.3 Repeated limits

In the previous section we saw that for the double limit,

$$\lim_{(x,y)\to(a,b)} f(x,y),$$

to exist f must have the same limiting value for *all* paths of approach to

(a, b). The limits along two particular paths are of special interest and lead to the *repeated limits*

$$L_1 = \lim_{x \to a} \lim_{y \to b} f(x,y) \quad \text{and} \quad L_2 = \lim_{y \to b} \lim_{x \to a} f(x,y).$$

That is

$$L_1 = \lim_{x \to a} g(x), \qquad \text{where } g(x) = \lim_{y \to b} f(x,y),$$

and

$$L_2 = \lim_{y \to b} h(y), \qquad \text{where } h(y) = \lim_{x \to a} f(x,y).$$

Of course, if the double limit exists, then so do these repeated limits and all three have the same value. However, the repeated limits may exist when the double limit does not. A more general definition of repeated limit is given in Exercises 10(a) No. 3.

Example 1 $f(x,y) = \dfrac{x-y}{x+y}$, $\quad (x,y) \neq (0,0)$.

Then

$$\lim_{x \to 0} \lim_{y \to 0} \left(\frac{x-y}{x+y}\right) = \lim_{x \to 0} (1) = 1,$$

and

$$\lim_{y \to 0} \lim_{x \to 0} \left(\frac{x-y}{x+y}\right) = \lim_{y \to 0} (-1) = -1.$$

Since the repeated limits are unequal, then the double limit cannot exist.

Example 2 $f(x,y) = \dfrac{xy^2}{x^2+y^4}$, $\quad (x,y) \neq (0,0)$.

As we saw in Example 3, Section 10.2, the double limit does not exist. But

$$\lim_{x \to 0} \lim_{y \to 0} f(x,y) = \lim_{x \to 0} (0/x^2) = 0$$

and

$$\lim_{y \to 0} \lim_{x \to 0} f(x,y) = 0.$$

Example 3 If

$$\lim_{(x,y) \to (a,b)} f(x,y) = l$$

and if

$$\lim_{y \to b} f(x,y) = g(x)$$

exists for every x sufficiently near to a, then

$$\lim_{x \to a} g(x) = l.$$

Since the double limit exists,

$$\forall\epsilon > 0; \exists\delta \,.\, |f(x,y) - l| < \tfrac{1}{2}\epsilon \qquad \text{for } 0 < \eta < \delta, \tag{1}$$

vhere

$$\eta^2 = (x-a)^2 + (y-b)^2.$$

.et x_1 be in $0 < \eta < \delta$ and sufficiently near to a for

$$\lim_{y \to b} f(x_1, y) = g(x_1).$$

That is,

$$\exists\delta_1 \,.\, |f(x_1,y) - g(x_1)| < \tfrac{1}{2}\epsilon \qquad \text{for } |y-b| < \delta_1. \tag{2}$$

Then if (x, y) satisfies both inequalities (1) and (2),

$$|g(x) - l| \leqslant |g(x) - f(x,y)| + |f(x,y) - l|$$

$$< \epsilon \qquad \text{for all } x \text{ such that } |x-a| < \delta.$$

10.4 Continuity

The function f is *continuous in* (x, y) at (a, b) if $f(a, b)$ is defined and

$$\forall\epsilon > 0; \exists\delta \,.\, |f(x,y) - f(a,b)| < \epsilon \qquad \text{for all } (x,y) \text{ in } 0 < \eta < \delta,$$

where $\eta^2 = (x-a)^2 + (y-b)^2$. In other words this requires that $f(x,y) \to f(a,b)$ as $(x,y) \to (a,b)$ by any path.

Example 1 $f(x,y) = (x+y)\sin(x+y)$.

As we saw in Example 1, Section 10.2, $f \to 0$ as $(x,y) \to (0,0)$; since $f(0,0) = 0$ we have shown that f is continuous in (x, y) at $(0, 0)$.

Example 2 $f(x,y) = \dfrac{xy}{x^2+y^2}, \quad x^2+y^2 \neq 0, \quad f(0,0) = 0.$

This function is not continuous in (x, y) at $(0, 0)$ (see Example 2, Section 10.2). However, at any other point $(a, b) \neq (0, 0)$ the function is continuous. Let $x = a + h, y = b + k$; then

$$f(a+h, b+k) - f(a,b) = \frac{(a^2 - b^2 + ah - bk)(ak - bh)}{(a^2+b^2)[(a+h)^2 + (b+k)^2]}.$$

If $|h| < \frac{1}{2}|a|$ and $|k| < \frac{1}{2}|b|$,

$$(a+h)^2 + (b+k)^2 > \tfrac{1}{4}(a^2+b^2)$$

and

$$|a^2 - b^2 + ah - bk| \leqslant a^2 + b^2 + |a|\,|h| + |b|\,|k|$$

$$< 3(a^2+b^2)/2.$$

Therefore

$$|f(a+h,b+k)-f(a,b)|<\frac{6(|ak|+|bh|)}{a^2+b^2}$$

$$<\epsilon \qquad \text{for } |h|, |k|<\delta,$$

where $\delta = \min(\delta_1, \delta_2)$, and

$$|h|<\delta_1 = \frac{(a^2+b^2)\epsilon}{12\,|b|} \quad \text{and} \quad |k|<\delta_2 = \frac{(a^2+b^2)\epsilon}{12\,|a|}.$$

Note. Continuity in (x, y) implies continuity in each variable separately, but the converse is not true. Thus Example 2 above is continuous in x when y is fixed and also continuous in y when x is held constant. However, as we have seen, f is not continuous in (x, y) at $(0, 0)$.

EXERCISES 10(a)

1. Find, when they exist, the repeated limits and the double limit of the given functions as $(x, y) \to (0, 0)$:

   (a) $\dfrac{x^2y^2}{x^4+y^4}$ (b) $\dfrac{x^2y^2}{x^2+y^2}$ (c) $\dfrac{x+y^2}{x^2+y}$

   (d) $x + y\sin(1/x)$ (e) $|x|^y$ (f) $\cot^{-1}[x^2+y^2)^{-1}]$.

2. Find, when they exist, the repeated limits and the double limit of each given function as $x \to \infty$ and $y \to \infty$:

   (a) $\dfrac{x}{x+y}$ (b) $\dfrac{xy}{x^2+y^2}$ (c) $\dfrac{x+y}{x^2+y^2}$.

3. A more general definition of repeated limit is

$$\lim_{x\to a}\lim_{y\to b} f(x,y) = \lim_{x\to a}\,[\,\overline{\lim_{y\to b}}\, f(x,y) - \underline{\lim_{y\to b}}\, f(x,y)],$$

   with a similar definition for the other repeated limit. Use this definition to find the repeated limits at the origin of the following functions:

   (a) $x + y\sin(1/x)$ (b) $x\sin(1/y) + y\sin(1/x)$

   (c) $(x+y)\sin\left(\dfrac{1}{x}+\dfrac{1}{y}\right)$.

4. Decide whether the given function $f(x, y)$ is continuous at the origin:

(a) $\sqrt{(x^2+y^2)}$ (b) $\dfrac{x^2y^2}{x^4+y^4}$, $(x,y)\neq(0,0)$, $f(0,0) = 0$

(c) $\dfrac{xy^2}{x^2+y^4}$, $(x,y)\neq(0,0)$, $f(0,0) = 0$

(d) $\dfrac{xy}{\sqrt{(x^2+y^2)}}$, $(x,y)\neq(0,0)$, $f(0,0) = 0$

(e) $\dfrac{(x-2y)(2x-y)}{x^2+y^2}$, $xy\neq 0$, $f(x,0) = f(0,y) = f(0,0) = 2$

(f) $\dfrac{x^2+y^2}{x-y}$, $x\neq y$, $f(x,y) = 0$ when $x = y$

(g) $(x+y)\sin\left(\dfrac{1}{x}+\dfrac{1}{y}\right)$, $xy\neq 0$, $f(x,0)=f(0,y)=f(0,0)=0$

(h) $xy\sin(x^{-1}y^{-1})$, $xy\neq 0$, $f(x,0)=f(0,y)=f(0,0)=0$

(i) $x\sin[4\tan^{-1}(y/x)]$, $x>0$, $f(0,y)=0$ for all y.

5. Show that $f(x,y)$ is continuous in (x,y) at $(0,0)$ if $f(r\cos\theta, r\sin\theta)$ is continuous at $r=0$ uniformly with respect to θ.

   Show that the converse is also true.

6. Prove that

$$f(x,y) = (x^4+y^4)/(x^2+y^2), \qquad (x,y)\neq(0,0),$$

   $f(0,0)=0$ is continuous in (x,y) at $(0,0)$.

7. Show that, if $f(x,y)$ is continuous in (x,y) at (a,b) and $f(a,b)>0$, then there is a neighbourhood of (a,b) in which $f(x,y)>0$.

10.5 Partial derivatives

If, in the function $f(x,y)$, y is held constant and the resulting function of x is differentiated with respect to x, we obtain the *partial derivative of f with respect to x*. This is written $\partial f/\partial x$ or f_x. Thus we define

$$\frac{\partial f}{\partial x} = \lim_{h\to 0}\frac{f(x+h,y)-f(x,y)}{h}$$

whenever the limit exists.

Similarly we define the partial derivative of f with respect to y to be

$$\frac{\partial f}{\partial y} = f_y = \lim_{k \to 0} \frac{f(x, y+k) - f(x, y)}{k}$$

when this limit exists.

When it is necessary to show which variable is being held constant we write respectively

$$\left.\frac{\partial f}{\partial x}\right|_y \quad \text{and} \quad \left.\frac{\partial f}{\partial y}\right|_x .$$

It will be noticed that the definition of a partial derivative corresponds closely to that of an ordinary derivative of a function of one variable. Consequently the *rules* for differentiation of a function of one variable are obeyed; this is illustrated in the examples below. Notice (Example 2) that the existence of the first partial derivatives does not guarantee the continuity of the function.

Example 1 $f(x, y) = x^2y^3 - x\cos y$.

Then $\dfrac{\partial f}{\partial x} = 2xy^3 - \cos y$ and $\dfrac{\partial f}{\partial y} = 3x^2y^2 + x\sin y$.

Example 2 $f(x, y) = \dfrac{xy}{x^2+y^2}$, $f(0, 0) = 0$.

Then

$$\frac{\partial f}{\partial x} = \frac{y(x^2+y^2) - 2x^2y}{(x^2+y^2)^2} = \frac{y(y^2-x^2)}{(x^2+y^2)^2}, \qquad (x, y) \neq (0, 0)$$

and

$$\frac{\partial f}{\partial y} = \frac{x(x^2+y^2) - 2xy^2}{(x^2+y^2)^2} = \frac{x(x^2-y^2)}{(x^2+y^2)^2}, \qquad (x, y) \neq (0, 0).$$

However, to find $f_x(0, 0)$, that is the value of the partial derivative of f with respect to x at the origin, we must return to the definition of a partial derivative. Thus

$$f_x(0, 0) = \lim_{h \to 0} \frac{f(h, 0) - f(0, 0)}{h} = 0.$$

Similarly $f_y(0, 0) = 0$. Notice that

$$\lim_{(x, y) \to (0, 0)} f_x(x, y)$$

does not exist. In other words the function

$$f_x(x, y) = \frac{y(y^2-x^2)}{(x^2+y^2)^2}$$

is not continuous in (x, y) at the origin.

Example 3 The two equations $x = r\cos\theta, y = r\sin\theta$ in the four variables x, y, r, θ allow us to express any two of the variables in terms of the other two. Thus $x = r\cos\theta, y = r\sin\theta$ give x and y in terms of r and θ, while

(a) $r^2 = x^2 + y^2, \theta = \tan^{-1}(y/x)$ gives r, θ in terms of x and y;
(b) $r = x\sec\theta, y = x\tan\theta$ gives r, y in terms of x and θ.

From the original equations we have

$$\left.\frac{\partial x}{\partial r}\right|_\theta = \cos\theta.$$

From (a),

$$2r\frac{\partial r}{\partial x} = 2x \quad \text{or} \quad \frac{\partial r}{\partial x} = \frac{x}{r} = \cos\theta,$$

and from (b)

$$\frac{\partial r}{\partial x} = \sec\theta.$$

The different results in these last two derivatives arise because different variables are being held constant in the two cases. In such circumstances it is essential to show which variable is being held constant. We write the two results

$$\left.\frac{\partial r}{\partial x}\right|_y = \cos\theta \quad \text{and} \quad \left.\frac{\partial r}{\partial x}\right|_\theta = \sec\theta.$$

Notice that $\partial r/\partial x$ is the reciprocal of $\partial x/\partial r$ only when the same variable is held constant in the two derivatives.

Example 4 $f(x, y, z) = xy^2z^3$.

The partial derivatives of functions of more than two variables are found by holding all save one variable constant and differentiating in the usual way with respect to the remaining variable. Thus

$$f_x = \left.\frac{\partial f}{\partial x}\right|_{y,z} = y^2z^3, \quad f_y = 2xyz^3, \quad f_z = 3xy^2z^2.$$

10.6 Differentiability

Definition The function $z(x, y)$ is differentiable at (x, y) if it is defined in a neighbourhood of (x, y) and

$$z(x + h, y + k) - z(x, y) = Ah + Bk + \rho\epsilon,$$

where $\rho = |h| + |k|$, $\epsilon \to 0$ as $\rho \to 0$ and A, B are independent of the increments h, k.

If it is more convenient we can take $\rho = \sqrt{(h^2 + k^2)}$.
When z is differentiable at (x, y), then taking $k = 0$,

$$\Delta z = z(x+h, y+k) - z(x, y) = Ah + \epsilon\,|h|,$$

whence $A = \partial z/\partial x$. Similarly $B = \partial z/\partial y$; thus differentiability implies the existence of the first partial derivatives and also the continuity of the function at the point. The converse is false.

Example $z(x, y) = \dfrac{x^3 - y^3}{x^2 + y^2}$, $\quad (x, y) \neq (0,.0); \quad z(0, 0) = 0.$

Then

$$z_x(0, 0) = \lim_{h \to 0} \frac{z(h, 0) - z(0, 0)}{h} = 1,$$

$$z_y(0, 0) = \lim_{k \to 0} \frac{z(0, k) - z(0, 0)}{k} = -1,$$

and

$$\Delta z = z(h, k) - z(0, 0) = (h^3 - k^3)/(h^2 + k^2).$$

Let $h = \rho \cos\theta$, $k = \rho \sin\theta$; then for differentiability at the origin

$$\rho(\cos^3\theta - \sin^3\theta) = A\rho\cos\theta + B\rho\sin\theta + \rho\epsilon,$$

where A, B are independent of h, k, $\epsilon \to 0$ as $\rho \to 0$ and we have taken $\rho^2 = h^2 + k^2$.

Cancelling ρ and letting $\rho \to 0$ gives

$$\cos^3\theta - \sin^3\theta = A\cos\theta + B\sin\theta,$$

which cannot be true for arbitrary θ. Therefore z is not differentiable at the origin.

10.7 Partial derivatives of second and higher orders

Since the partial derivatives of a function $f(x, y)$ are themselves functions of x and y, they, in turn, may possess partial derivatives. These *second-order partial derivatives* are defined as follows:

$$\frac{\partial^2 f}{\partial x^2} = f_{xx}(x, y) = \lim_{h \to 0} \frac{f_x(x+h, y) - f_x(x, y)}{h}$$

$$\frac{\partial^2 f}{\partial y^2} = f_{yy}(x,y) = \lim_{k \to 0} \frac{f_y(x,y+k) - f_y(x,y)}{k}$$

when these limits exist.

In addition there are two 'mixed' second derivatives

$$\frac{\partial^2 f}{\partial x\, \partial y} = \frac{\partial}{\partial x}\left(\frac{\partial f}{\partial y}\right) = f_{xy}(x,y) = \lim_{h \to 0} \frac{f_y(x+h,y) - f_y(x,y)}{h}$$

$$= \lim_{h \to 0} \left(\lim_{k \to 0} \frac{f(x+h,y+k) - f(x+h,y)}{k} - \lim_{k \to 0} \frac{f(x,y+k) - f(x,y)}{k} \right)$$

$$= \lim_{h \to 0} \lim_{k \to 0} \frac{f(x+h,y+k) - f(x+h,y) - f(x,y+k) + f(x,y)}{hk}$$

$$= \lim_{h \to 0} \lim_{k \to 0} \frac{\Delta^2 f}{hk}.$$

Also

$$\frac{\partial^2 f}{\partial y\, \partial x} = \frac{\partial}{\partial y}\left(\frac{\partial f}{\partial x}\right) = f_{yx}(x,y) = \lim_{k \to 0} \frac{f_x(x,y+k) - f_x(x,y)}{k}$$

$$= \lim_{k \to 0} \lim_{h \to 0} \frac{\Delta^2 f}{hk}.$$

In other words the two 'mixed' second derivatives are the two repeated limits of $\Delta^2 f/kh$. They are equal only when these repeated limits are equal. In the next section we consider sufficient conditions for their equality. However, in most of the elementary functions we meet the order of differentiation does not matter.

Note: In the definitions of the 'mixed' second-order derivatives we have obtained f_{xy} from $\partial^2 f/\partial x\, \partial y$ by suppressing the ∂'s. Some authors write f_{xy} for $(f_x)_y$ by suppressing the bracket; in our notation this is $f_{yx} = \partial^2 f/\partial y\, \partial x$.

Third-order derivatives

$$\frac{\partial^3 f}{\partial x^3}, \quad \frac{\partial^3 f}{\partial y\, \partial x^2}, \quad \frac{\partial^3 f}{\partial x^2\, \partial y}, \quad \frac{\partial^3 f}{\partial x\, \partial y^2}, \quad \frac{\partial^3 f}{\partial y^2\, \partial x}, \quad \frac{\partial^3 f}{\partial y^3}$$

are defined in the obvious way; 'mixed' derivatives such as

$$\frac{\partial^3 f}{\partial y\,\partial x^2} \quad \text{and} \quad \frac{\partial^3 f}{\partial x^2\,\partial y}$$

may or may not be equal.

Example 1 If $z = e^{ax}\cos by$, then

$$z_x = ae^{ax}\cos by, \quad z_y = -be^{ax}\sin by$$

$$z_{xx} = a^2 e^{ax}\cos by, \quad z_{yy} = -b^2 e^{ax}\cos by$$

$$z_{xy} = -abe^{ax}\sin by = z_{yx}.$$

Example 2 If $u = z\tan^{-1}(y/x), x \neq 0$, then

$$\frac{\partial u}{\partial x} = \frac{-yz}{x^2+y^2}, \quad \frac{\partial u}{\partial y} = \frac{xz}{x^2+y^2}, \quad \frac{\partial u}{\partial z} = \tan^{-1}\left(\frac{y}{x}\right)$$

$$\frac{\partial^2 u}{\partial x^2} = \frac{2xyz}{(x^2+y^2)^2}, \quad \frac{\partial^2 u}{\partial y^2} = \frac{-2xyz}{(x^2+y^2)^2}, \quad \frac{\partial^2 u}{\partial z^2} = 0$$

$$\frac{\partial^2 u}{\partial x\,\partial y} = \frac{z(y^2-x^2)}{(x^2+y^2)^2} = \frac{\partial^2 u}{\partial y\,\partial x},$$

$$\frac{\partial^2 u}{\partial z\,\partial x} = \frac{-y}{x^2+y^2} = \frac{\partial^2 u}{\partial x\,\partial z}.$$

Example 3 If

$$f(x,y) = \frac{xy(x^2-y^2)}{(x^2+y^2)}, \qquad x^2+y^2 \neq 0, \quad f(0,0) = 0,$$

then $f_{xy}(0,0) \neq f_{yx}(0,0)$.

$$f_x(0,k) = \lim_{h\to 0}\frac{f(h,k)-f(0,k)}{h}$$

$$= \lim_{h\to 0}\frac{k(h^2-k^2)}{h^2+k^2} = -k, \qquad k \neq 0$$

and $f_x(0,0) = 0$. Also

$$f_y(h,0) = \lim_{k\to 0}\frac{f(h,k)-f(h,0)}{k} = h, \quad h \neq 0, \quad f_y(0,0) = 0.$$

Then

$$f_{xy}(0,0) = \lim_{h\to 0} \frac{f_y(h,0)-f_y(0,0)}{h} = 1,$$

and

$$f_{yx}(0,0) = \lim_{k\to 0} \frac{f_x(0,k)-f_x(0,0)}{k} = -1.$$

10.8 Equality of 'mixed' derivatives

Theorem If f_x, f_y and f_{yx} all exist in a neighbourhood of (a, b) and f_{yx} is continuous in (x, y) at (a, b), then $f_{xy}(a, b)$ also exists and $f_{xy}(a, b) = f_{yx}(a, b)$.

Proof Let $\phi(x) = f(x, y+k) - f(x, y)$, then

$$\begin{aligned}\Delta^2 f &= \phi(x+h)-\phi(x)\\ &= h\phi'(x+\theta_1 h) \qquad (0<\theta_1<1) \quad \text{(using the mean value theorem (Section 4.8))}\\ &= h[f_x(x+\theta_1 h, y+k) - f_x(x+\theta_1 h, y)].\end{aligned}$$

Using the mean value theorem on this function of y, we have near (a, b)

$$\Delta^2 f = hkf_{yx}(a+\theta_1 h, b+\theta_2 k), \qquad 0<\theta_2<1.$$

Since f_{yx} is continuous at (a, b),

$$\Delta^2 f = hk[f_{yx}(a,b)+\epsilon], \qquad \text{where } \epsilon \to 0 \text{ as } \rho = |h|+|k| \to 0.$$

Then, if it exists,

$$\begin{aligned} f_{xy}(a,b) &= \lim_{h\to 0}\lim_{k\to 0} \frac{\Delta^2 f}{hk},\\ &= \lim_{h\to 0}\lim_{k\to 0} [f_{yx}(a,b)+\epsilon]\\ &= f_{yx}(a,b),\end{aligned}$$

and the theorem is proved.

Another theorem, which requires the existence of all the second-order derivatives, is the following.

Theorem If f_x and f_y exist in a neighbourhood of (a, b) and are differentiable at (a, b), then $f_{xy}(a, b) = f_{yx}(a, b)$.

Proof This is left as an exercise.

EXERCISES 10(b)

1. Find $\partial z/\partial x$ and $\partial z/\partial y$ for the following functions $z = z(x, y)$:

(a) $\exp(x - 2y)$ (b) $\log(x - y^2)$ (c) $\tan(x/y)$

(d) $y(1 - x^2y^2)^{-1/2}$ (e) $\text{ch}^2(x^2 - y^2)$ (f) $e^x(x\cos y - y\sin y)$

(g) $\dfrac{x-y}{x+y}$ (h) $\text{th}^{-1}\left(\dfrac{x-y}{x+y}\right)$ (i) $e^{xy^2}\sin x^2y$.

2. Given $x = r\cos\theta$ and $y = r\sin\theta$, prove that

(a) $\left.\dfrac{\partial r}{\partial \theta}\right|_y = -r\cot\theta$ (b) $\left.\dfrac{\partial y}{\partial r}\right|_x \left.\dfrac{\partial y}{\partial r}\right|_\theta = 1$

(c) $\left.\dfrac{\partial x}{\partial r}\right|_y \left.\dfrac{\partial r}{\partial y}\right|_x + r\left.\dfrac{\partial \theta}{\partial r}\right|_y = 0.$

3. Find, when they exist, $f_x(0, 0)$ and $f_y(0, 0)$ for the following functions $f(x, y)$:

(a) $\dfrac{xy^2}{x^2+y^2}$ (b) $x\sin\left(\dfrac{1}{x^2+y^2}\right)$ (c) $\dfrac{xy}{\sqrt{(x^2+y^2)}}$.

In each case assume $f(0, 0) = 0$. Investigate the continuity of the derivatives at the origin.

4. Show that $f(x, y) = |xy|^k$ is differentiable at the origin if $k > \frac{1}{2}$.

5. If $f(x, y) = x\sin[4\tan^{-1}(y/x)]$, $x > 0$; $f(0, y) = 0$ for all y, find which first and second derivatives exist at the origin.

6. Show that if $z(x, y) = f(x) + f(y)$, where

$$f(x) = x^2\sin(1/x), \qquad x \neq 0, \quad f(0) = 0,$$

then $z_x(0, 0) = 0 = z_y(0, 0)$, and z is differentiable at the origin.

7. Given that f_x and f_y are bounded in a neighbourhood of (a, b), prove that $f(x, y)$ is a continuous function of (x, y) at (a, b).

8. Prove that if f_x and f_y are continuous functions of (x, y) at (a, b), then f is differentiable at (a, b).

9. Find the second derivatives of the functions in 1(a), (b) and (c) above.

10. Show that if

$$f(x, y) = \frac{xy(x^2 - y^2)}{x^2 + y^2}, \quad f(0, 0) = 0,$$

then $f_{xy}(0, 0) \neq f_{yx}(0, 0)$.

11. Given that

$$f(x, y) = x^2 \tan^{-1}(y/x) - y^2 \tan^{-1}(x/y), \qquad xy \neq 0,$$

$$f(x, 0) = f(0, y) = f(0, 0) = 0,$$

show that $f_{xy}(0, 0) \neq f_{yx}(0, 0)$.

12. Prove that if f_{xy} and f_{yx} are continuous in a neighbourhood of (a, b) then $f_{xy}(a, b) = f_{yx}(a, b)$.

13. Prove that if f_x and f_y exist in a neighbourhood of (a, b), and are differentiable at (a, b), then $f_{xy}(a, b) = f_{yx}(a, b)$.

10.9 Differentials

We have seen that if a function $z(x, y)$ is differentiable at a point (a, b), then

$$\Delta z = z(x + h, y + k) - z(x, y) = h\frac{\partial z}{\partial x} + k\frac{\partial z}{\partial y} + \rho\epsilon,$$

where $\rho = |h| + |k|$ and $\epsilon \to 0$ as $\rho \to 0$. We define the *differential* of z, $\mathrm{d}z$, to be the principal part of Δz and write

$$\Delta z = \mathrm{d}z + \rho\epsilon.$$

Further, we define the differentials of the independent variables $\mathrm{d}x$, $\mathrm{d}y$, to be the same as the increments in those variables, that is

$$\mathrm{d}z = \frac{\partial z}{\partial x}\,\mathrm{d}x + \frac{\partial z}{\partial y}\,\mathrm{d}y.$$

This is called the *total differential* of z.

The following theorem shows that the total differential of a function always has the same form *whether or not the variables are independent*.

Theorem If $u = u(x, y)$ is a differentiable function of (x, y) and x, y are differentiable functions of the independent variables s, t, then u considered as a function of s, t is differentiable and its differential is given by

$$du = \frac{\partial u}{\partial x}\,dx + \frac{\partial u}{\partial y}\,dy.$$

Proof Since x, y are differentiable functions of s and t, and s, t are independent variables,

$$\begin{aligned}\Delta x &= \frac{\partial x}{\partial s}\,ds + \frac{\partial x}{\partial t}\,dt + \epsilon_1\rho' = dx + \epsilon_1\rho' \\ \Delta y &= \frac{\partial y}{\partial s}\,ds + \frac{\partial y}{\partial t}\,dt + \epsilon_2\rho' = dy + \epsilon_2\rho'\end{aligned} \tag{1}$$

where $\rho' = |ds| + |dt|$ and $\epsilon_1, \epsilon_2 \to 0$ as $\rho' \to 0$. Then, if

$$\eta = \max(\epsilon_1, \epsilon_2) \quad \text{and} \quad M = \max(|x_s|, |x_t|, |y_s|, |y_t|),$$

$$|\Delta x| < (M+\eta)\rho' \quad \text{and} \quad |\Delta y| < (M+\eta)\rho';$$

also

$$\frac{|\Delta x| + |\Delta y|}{\rho'} < 2(M+\eta).$$

Therefore

$$\rho = |\Delta x| + |\Delta y| \to 0 \quad \text{as } \rho' \to 0.$$

Since $u(x, y)$ is a differentiable function of x and y,

$$\Delta u = \frac{\partial u}{\partial x}\,\Delta x + \frac{\partial u}{\partial y}\,\Delta y + \epsilon\rho,$$

where $\epsilon \to 0$ as $\rho = |\Delta x| + |\Delta y| \to 0$. Using eqs. (1)

$$\begin{aligned}\Delta u &= \frac{\partial u}{\partial x}\,dx + \frac{\partial u}{\partial y}\,dy + \left(\epsilon\rho + \epsilon_1\rho'\frac{\partial u}{\partial x} + \epsilon_2\rho'\frac{\partial u}{\partial y}\right) \\ &= \frac{\partial u}{\partial x}\,dx + \frac{\partial u}{\partial y}\,dy + \epsilon'\rho',\end{aligned}$$

where

$$\epsilon' = \epsilon\frac{\rho}{\rho'} + \epsilon_1\frac{\partial u}{\partial x} + \epsilon_2\frac{\partial u}{\partial y}.$$

Since the coefficients of $\epsilon, \epsilon_1, \epsilon_2$ are bounded, $\epsilon' \to 0$ as $\rho' \to 0$. Hence

$$du = \frac{\partial u}{\partial x}\,dx + \frac{\partial u}{\partial y}\,dy.$$

Example 1 If $x = r\cos\theta, y = r\sin\theta$, then

$$\begin{aligned}dx &= \cos\theta\,dr - r\sin\theta\,d\theta, \\ dy &= \sin\theta\,dr + r\cos\theta\,d\theta.\end{aligned}$$

'he above theorem shows that if any one of the four differentials, dr say), is found in terms of two others, dy, dθ (say), then the coefficients f the latter are the partial derivatives of r with respect to y and θ. Thus

$$\mathrm{d}r = \operatorname{cosec}\theta\, \mathrm{d}y - r\cot\theta\, \mathrm{d}\theta;$$

which implies that

$$\left.\frac{\partial r}{\partial y}\right|_{\theta} = \operatorname{cosec}\theta \quad \text{and} \quad \left.\frac{\partial r}{\partial \theta}\right|_{y} = -r\cot\theta.$$

xample 2 If $x = u^2 + v^2, y = u^2 - v^2$, show that

$$\left.\frac{\partial x}{\partial u}\right|_{v} \left.\frac{\partial u}{\partial x}\right|_{y} = \frac{1}{2}.$$

Iere d$x = 2u$ d$u + 2v$ dv and d$y = 2u$ d$u - 2v$ dv. Eliminating dv, we have

$$\mathrm{d}u = \frac{1}{4u}(\mathrm{d}x + \mathrm{d}y).$$

'herefore

$$\left.\frac{\partial u}{\partial x}\right|_{y} = \frac{1}{4u}.$$

'rom $x = u^2 + v^2$, we have $(\partial x/\partial u)|_v = 2u$; hence the result.

From the above theorem we deduce

'orollary 1 If $u(x, y)$ is a differentiable function of x and y, which are ifferentiable functions of a single independent variable t, then

$$\frac{\mathrm{d}u}{\mathrm{d}t} = \frac{\partial u}{\partial x}\frac{\mathrm{d}x}{\mathrm{d}t} + \frac{\partial u}{\partial y}\frac{\mathrm{d}y}{\mathrm{d}t}.$$

'orollary 2 If x, y are differentiable functions of variables s, t, then

$$\frac{\partial u}{\partial s} = \frac{\partial u}{\partial x}\frac{\partial x}{\partial s} + \frac{\partial u}{\partial y}\frac{\partial y}{\partial s},$$

nd

$$\frac{\partial u}{\partial t} = \frac{\partial u}{\partial x}\frac{\partial x}{\partial t} + \frac{\partial u}{\partial y}\frac{\partial y}{\partial t}.$$

:xample 1 If $u = \log(x^2 + y^2)$ and $x = e^t, y = e^{-t}$, then

$$\frac{\mathrm{d}u}{\mathrm{d}t} = \frac{2xe^t}{x^2 + y^2} - \frac{2ye^{-t}}{x^2 + y^2} = 2\ \mathrm{th}\ 2t.$$

Example 2 If $u = x^2 + y^2 + z^2$, $x = r\cos\theta$, $y = r\sin\theta$ and $z = r$, then

$$\frac{\partial u}{\partial r} = 2x\cos\theta + 2y\sin\theta + 2z = 4r,$$

and

$$\frac{\partial u}{\partial \theta} = -2xr\sin\theta + 2yr\cos\theta = 0.$$

Example 3 If $x = r\cos\theta$, $y = r\sin\theta$ and V is a differentiable function, show that

$$\frac{\partial^2 V}{\partial x^2} + \frac{\partial^2 V}{\partial y^2} = \frac{\partial^2 V}{\partial r^2} + \frac{1}{r}\frac{\partial V}{\partial r} + \frac{1}{r^2}\frac{\partial^2 V}{\partial \theta^2}.$$

We have

$$\frac{\partial V}{\partial r} = \frac{\partial V}{\partial x}\cos\theta + \frac{\partial V}{\partial y}\sin\theta,$$

and

$$\frac{\partial V}{\partial \theta} = -r\frac{\partial V}{\partial x}\sin\theta + r\frac{\partial V}{\partial y}\cos\theta.$$

Solving these equations for $\partial V/\partial x$ and $\partial V/\partial y$, we find

$$\frac{\partial V}{\partial x} = \frac{\partial V}{\partial r}\cos\theta - \frac{1}{r}\frac{\partial V}{\partial \theta}\sin\theta,$$

$$\frac{\partial V}{\partial y} = \frac{\partial V}{\partial r}\sin\theta + \frac{1}{r}\frac{\partial V}{\partial \theta}\cos\theta.$$

Then

$$\frac{\partial^2 V}{\partial x^2} = \left(\cos\theta\frac{\partial}{\partial r} - \frac{1}{r}\sin\theta\frac{\partial}{\partial \theta}\right)\left(\cos\theta\frac{\partial V}{\partial r} - \frac{1}{r}\sin\theta\frac{\partial V}{\partial \theta}\right)$$

$$= \cos^2\theta\frac{\partial^2 V}{\partial r^2} + \frac{1}{r}\sin^2\theta\frac{\partial V}{\partial r} + \frac{1}{r^2}\sin^2\theta\frac{\partial^2 V}{\partial \theta^2}$$

$$+ 2\sin\theta\cos\theta\left(\frac{1}{r^2}\frac{\partial V}{\partial \theta} - \frac{1}{r}\frac{\partial^2 V}{\partial r\,\partial \theta}\right).$$

Similarly

$$\frac{\partial^2 V}{\partial y^2} = \left(\sin\theta\frac{\partial}{\partial r} + \frac{1}{r}\cos\theta\frac{\partial}{\partial \theta}\right)\left(\sin\theta\frac{\partial V}{\partial r} + \frac{1}{r}\cos\theta\frac{\partial V}{\partial \theta}\right)$$

$$= \sin^2\theta\frac{\partial^2 V}{\partial r^2} + \frac{1}{r}\cos^2\theta\frac{\partial V}{\partial r} + \frac{1}{r^2}\cos^2\theta\frac{\partial^2 V}{\partial \theta^2}$$

$$- 2\sin\theta\cos\theta\left(\frac{1}{r^2}\frac{\partial V}{\partial \theta} - \frac{1}{r}\frac{\partial^2 V}{\partial r\,\partial \theta}\right).$$

The required result then follows on addition.

Example 4 (*Euler's theorem*) A function $f(x, y)$ is homogeneous of degree n in x and y if and only if

$$x\frac{\partial f}{\partial x}+y\frac{\partial f}{\partial y} = nf.$$

$f(x, y)$ is said to be *homogeneous* of degree n in x and y if

$$f(\lambda x, \lambda y) = \lambda^n f(x, y) \qquad \text{for all } x \text{ and } y.$$

For example, x^2+xy+y^2 is homogeneous of degree 2; $\tan^{-1}(x/y)$ is homogeneous of degree 0; $(x^2+4y^2-z^2)^{-1/3}$ is homogeneous of degree $-\frac{2}{3}$ in x, y and z.

Let $\xi=\lambda x, \eta=\lambda y$; then $f(\xi, \eta)=\lambda^n f(x, y)$. Differentiating with respect to λ, we obtain

$$x\frac{\partial f}{\partial \xi}+y\frac{\partial f}{\partial \eta} = n\lambda^{n-1}f;$$

then putting $\lambda=1$ gives

$$xf_x+yf_y = nf.$$

Now suppose $xf_x+yf_y=nf$. Then

$$\frac{\mathrm{d}}{\mathrm{d}\lambda}f(\xi, \eta) = xf_\xi+yf_\eta = \frac{1}{\lambda}(\xi f_\xi+\eta f_\eta) = \frac{nf}{\lambda}.$$

Writing $u=f(\xi, \eta)$, we have, for all values of λ,

$$\frac{1}{u}\frac{\mathrm{d}u}{\mathrm{d}\lambda} = \frac{n}{\lambda}.$$

Hence $u=A\lambda^n$, where A is independent of λ. That is $f(\lambda x, \lambda y)=A\lambda^n$, and putting $\lambda=1$ gives $A=f(x, y)$. Therefore $f(\lambda x, \lambda y)=\lambda^n f(x, y)$ for all values of x and y.

EXERCISES 10(c)

1. Given $V=re^x+xe^r$, where $r^2=x^2+y^2$, show that

   (a) $yV_x-xV_y=y(re^x+e^r)$ (b) $xV_x+yV_y=V+xr(e^x+e^r)$.

2. Given $(x+y)z=f(x)+f(y)$, show that

$$(x+y)(z_x-z_y) = f'(x)-f'(y).$$

3. Find $\mathrm{d}z/\mathrm{d}t$ when $z=e^{xy^2}$, $x=t\cos t$ and $y=t\sin t$.

4. Given $z = z(x, y)$ and $u = x^2 - y^2 - 2xy$, $v = y$, transform the equation

$$(x+y)\frac{\partial z}{\partial x} + (x-y)\frac{\partial z}{\partial y} = 0$$

into one in terms of u and v. Hence solve the original equation.

5. Given $x = r\cos\theta$, $y = r\sin\theta$ where r, θ are functions of t, express $\dot{x}, \dot{y}$ in terms of $r, \theta, \dot{r}, \dot{\theta}$ and prove that

(a) $\ddot{x}\cos\theta + \ddot{y}\sin\theta = \ddot{r} - r\dot{\theta}^2$

(b) $\ddot{y}\cos\theta - \ddot{x}\sin\theta = \dfrac{1}{r}\dfrac{\mathrm{d}}{\mathrm{d}t}(r^2\dot{\theta})$.

6. Prove that if x, y, z are connected by a single relation $f(x, y, z) = 0$, then

$$\left.\frac{\partial z}{\partial y}\right|_x = -\frac{f_y}{f_z}.$$

Prove also that

$$\left.\frac{\partial z}{\partial y}\right|_x \left.\frac{\partial y}{\partial x}\right|_z = -\left.\frac{\partial z}{\partial x}\right|_y.$$

7. Show that if x, y, z satisfy the relations $f(x, y, z) =$ constant and $xyz =$ constant, then

$$\frac{\mathrm{d}y}{\mathrm{d}x} = \frac{-y(xf_x - zf_z)}{x(yf_y - zf_z)}.$$

8. Each variable x, y, z is a function of u and v. Prove that

$$\left.\frac{\partial x}{\partial u}\right|_y = \frac{\partial(x,y)}{\partial(u,v)}\bigg/\frac{\partial y}{\partial v} \quad \text{and} \quad \left.\frac{\partial y}{\partial x}\right|_z = \frac{\partial(y,z)}{\partial(u,v)}\bigg/\frac{\partial(x,z)}{\partial(u,v)},$$

where

$$\frac{\partial(x,y)}{\partial(u,v)} = \begin{vmatrix} x_u & y_u \\ x_v & y_v \end{vmatrix}$$

is called the *Jacobian* of x, y with respect to u, v.

9. Show that if $V = x^n f(y/x)$, then $xV_x + yV_y = nV$. Show that if $V_{xy} = 0$ and $u = y/x$, then

$$(n-1)\frac{\mathrm{d}f}{\mathrm{d}u} = u\frac{\mathrm{d}^2 f}{\mathrm{d}u^2}.$$

Hence show that V is of the form $ax^n + by^n$, where a, b are constants.

10. Show that if $f(x, y)$ is homogeneous of degree n in x and y, and twice differentiable, then

$$x^2 f_{xx} + 2xy f_{xy} + y^2 f_{yy} = n(n-1) f.$$

Show that if $V = \phi(u)$, where u is a homogeneous function of degree n in x and y, then

$$x^2 V_{xx} + 2xy V_{xy} + y^2 V_{yy} = n(n-1) u\phi'(u) + n^2 u^2 \phi''(u).$$

11. Using the substitutions $u = x - ct$, $v = x + ct$, transform the equation

$$\frac{\partial^2 y}{\partial t^2} = c^2 \frac{\partial^2 y}{\partial x^2}$$

into an equation in u and v. Hence solve the original equation.

12. In the equation

$$\frac{\partial v}{\partial t} = k \frac{\partial^2 v}{\partial x^2},$$

k is a constant. Assuming $v = f(u)$, where $u = x/\sqrt{t}$, show that f satisfies the equation $2kf'' + uf' = 0$. Solve this equation.

10.10 Taylor's theorem

Theorem If $f(x, y)$ and its partial derivatives of order n are continuous in a neighbourhood of (a, b) and $(a + h, b + k)$ is in this neighbourhood, then

$$\begin{aligned} f(a+h, b+k) &= f(a, b) + \left(h\frac{\partial}{\partial a} + \frac{\partial}{\partial b}\right) f(a, b) \\ &\quad + \frac{1}{2!}\left(h^2 \frac{\partial^2}{\partial a^2} + 2hk \frac{\partial^2}{\partial a\, \partial b} + k^2 \frac{\partial^2}{\partial b^2}\right) f(a, b) + \cdots \\ &\quad + \frac{1}{(n-1)!}\left(h\frac{\partial}{\partial a} + k\frac{\partial}{\partial b}\right)^{n-1} f(a, b) + R_n, \end{aligned}$$

where

$$R_n = \frac{1}{n!}\left(h\frac{\partial}{\partial x} + k\frac{\partial}{\partial y}\right)^n f(x, y)$$

evaluated at $(a + \theta h, b + \theta k)$, $0 < \theta < 1$.

Note

$$\left(h\frac{\partial}{\partial a}+k\frac{\partial}{\partial b}\right)f(a,b)$$

means $hf_x + kf_y$ evaluated at (a, b), and

$$\left(h\frac{\partial}{\partial a}+k\frac{\partial}{\partial b}\right)^n f(a,b)$$

stands for

$$\sum_{r=0}^{n}\binom{n}{r}h^r k^{n-r}\frac{\partial^n f(x,y)}{\partial x^r \partial y^{n-r}}$$

evaluated at (a, b).

Proof We apply Maclaurin's theorem (Section 8.1) to the function

$$F(t) = f(a+ht, b+kt) \text{ in the interval } 0 \leqslant t \leqslant 1.$$

Then

$$F(1) = F(0)+F'(0)+\frac{1}{2!}F''(0)+\cdots+\frac{1}{(n-1)!}F^{(n-1)}(0)$$
$$+\frac{1}{n!}F^{(n)}(\theta), \qquad 0<\theta<1.$$

Now

$$F'(t) = hf_x(a+ht, b+kt)+kf_y(a+ht, b+kt)$$

and so

$$F'(0) = hf_x(a,b)+kf_y(a,b) = \left(h\frac{\partial}{\partial a}+k\frac{\partial}{\partial b}\right)f(a,b).$$

Similarly

$$F''(0) = \left(h^2\frac{\partial^2}{\partial a^2}+2hk\frac{\partial^2}{\partial a\,\partial b}+k^2\frac{\partial^2}{\partial b^2}\right)f(a,b)$$
$$= \left(h\frac{\partial}{\partial a}+k\frac{\partial}{\partial b}\right)^2 f(a,b).$$

In particular, when $n = 1$, we have the *mean value theorem* for functions of two variables:

Theorem If $f(x, y)$ is continuous and differentiable in a neighbourhood of (a, b), then there exists a number $\theta, 0 < \theta < 1$, such that

$$f(a+h, b+k)-f(a,b) = f_x(a+\theta h, b+\theta k)+kf_y(a+\theta h, b+\theta k).$$

10.11 Maxima and minima

A function $f(x, y)$ has an *extreme value* at (a, b) if

$$\Delta f = f(a+h, b+k) - f(a, b)$$

is of constant sign throughout a neighbourhood of (a, b). The extreme value is a *minimum* if $\Delta f > 0$ and a *maximum* if $\Delta f < 0$ in the neighbourhood.

Clearly, a *necessary* condition for a differentiable function $f(x, y)$ to have an extreme value at (a, b) is

$$f_x(a, b) = 0 = f_y(a, b)$$

(compare with Section 4.6). For $f(a, b)$ cannot be an extreme value of $f(x, y)$ unless it is an extreme value of the function $f(x, b)$ and also of the function $f(a, y)$. A point at which $f_x = 0 = f_y$ is called a *stationary point* of the function.

This condition can be expressed in terms of differentials;

$$\mathrm{d}f = f_x(a, b)\,\mathrm{d}x + f_y(a, b)\,\mathrm{d}y$$

(see Section 10.9) at an extreme value. If x and y are not independent variables this relation is still true, but does not then imply that f_x and f_y both vanish. The condition is not sufficient. For example $z = xy$ represents a hyperbolic paraboloid; $z > 0$ in the first and third quadrants and $z < 0$ in the second and fourth. Consequently z does not have an extreme value at the origin although $z_x = z_y = 0$ there.

To find a sufficient condition for an extreme value we must investigate the second-order derivatives. Suppose $f(x, y)$ has continuous second derivatives in a neighbourhood of (a, b) and that $f_x(a, b) = 0 = f_y(a, b)$. Let $D = f_{xx}f_{yy} - f_{xy}^2$; then if $D > 0$ at (a, b), f has an extreme value at (a, b) which is a minimum or maximum according as $f_x(a, b)$ (and $f_{yy}(a, b)$) is positive or negative. If $D < 0$, f has a *saddle point* at (a, b). By Taylor's theorem

$$\Delta f = \tfrac{1}{2}[h^2 f_{xx}(a+\theta h, b+\theta k) + 2hk f_{xy}(a+\theta h, b+\theta k) + k^2 f_{yy}(a+\theta h, b+\theta k)], \qquad 0 < \theta < 1.$$

Since the second derivatives are continuous we can write

$$f_{xx}(a+\theta h, b+\theta k) = f_{xx}(a, b) + \epsilon_1,$$

where $\epsilon_1 \to 0$ as $\rho = |h| + |k| \to 0$, and similar expressions for f_{xy} and f_{yy}. That is

$$\Delta f = \tfrac{1}{2}(Ah^2 + 2Bhk + Ck^2) + \epsilon_1 h^2 + 2\epsilon_2 hk + \epsilon_3 k^2,$$

where

$$A = f_{xx}(a,b), \quad B = f_{xy}(a,b), \quad C = f_{yy}(a,b)$$

and $\epsilon_1, \epsilon_2, \epsilon_3 \to 0$ as $\rho \to 0$. Let $h = \rho \cos\theta, k = \rho \sin\theta$; then

$$f = \tfrac{1}{2}\rho^2 \{A \cos^2\theta + 2B \sin\theta \cos\theta + C \sin^2\theta + \epsilon\},$$

where $\epsilon = \epsilon_1 \cos^2\theta + 2\epsilon_2 \sin\theta \cos\theta + \epsilon_3 \sin^2\theta \to 0$ as $\rho \to 0$; note that ϵ is not necessarily positive. Then the sign of Δf depends on that of $E + \epsilon$, where

$$E = A \cos^2\theta + 2B \sin\theta \cos\theta + C \sin^2\theta.$$

First, suppose A does not vanish in the neighbourhood of (a, b). Then

$$E = [(A \cos\theta + B \sin\theta)^2 + (AC - B^2) \sin^2\theta]/A, \qquad A \neq 0,$$

and E keeps a constant sign and has a positive lower bound, $|E| \geqslant m$. Therefore, when ρ is sufficiently small, $|\epsilon| < m$ and Δf has the same sign as E.

If $AC - B^2 > 0$, then E has the same sign as A. That is f has a maximum at (a, b) if $A < 0$ and a minimum if $A > 0$.

If $AC - B^2 < 0$, then E changes sign; for AE is positive when $\sin\theta = 0$ and negative when $\tan\theta = -A/B$. In this case there is no extreme value, the point (a, b) being a saddle point (minimax) of f.

If $AC - B^2 = 0$, then AE is a perfect square and may vanish without changing sign; the sign of Δf then depends on that of ϵ. Further investigation is required in this 'doubtful case'.

If $A = 0, E = (2B \cos\theta + C \sin\theta) \sin\theta$, and if $B \neq 0, E$ changes sign as θ varies; again f has a saddle point at (a, b).

If $A = 0 = B, E = C \sin^2\theta$, again E can vanish without changing sign and so is the doubtful case.

Summarising, we have the following.

Theorem (a) $f(a, b)$ is an extreme value of $f(x, y)$ if

$$f_x(a,b) = 0 = f_y(a,b),$$

and

$$f_{xx}(a,b)\, f_{yy}(a,b) - f_{xy}^2(a,b) > 0;$$

the value is a minimum or maximum according as $f_{xx}(a, b)$ (or $f_{yy}(a, b)$) is positive or negative respectively.

(b) If $f_x(a, b) = 0 = f_y(a, b)$ and $f_{xx}f_{yy} - f_{xy}^2 < 0$ at (a, b), then f has a saddle point at (a, b).

(c) If $f_x(a, b) = 0 = f_y(a, b)$ and $f_{xx}f_{yy} - f_{xy}^2 = 0$ at (a, b), nothing can be inferred. (The 'doubtful case'.)

Example 1 $f(x, y) = x^4 - y^4 - 2(x^2 - y^2)$.

Then $f_x = 4x(x^2 - 1)$ and $f_y = 4y(1 - y^2)$. Hence $f_x = 0 = f_y$ when $x = 0, \pm 1$ and $y = 0, \pm 1$. Also

$$f_{xx} = 12x^2 - 4, \quad f_{yy} = 4 - 12y^2 \text{ and } f_{xy} = 0.$$

Table 10.1

	f_{xx}	f_{yy}	f_{xy}	$f_{xx}f_{yy} - f_{xy}^2$
(0, 1) (0, − 1)	−4	−8	0	32
(1, 0) (− 1, 0)	8	4	0	32
(0, 0)	−4	4	0	−16
(1, 1) (1, − 1) (− 1, 1) (− 1, −1)	8	−8	0	−64

Data for the nine points at which $f_x = f_y = 0$ are given in Table 10.1. From this, we see that f has a minimum value of − 1 at (± 1, 0) and a maximum value of 1 at (0, ± 1); f has saddle points at (0, 0) and (±1, ±1).

Example 2 $f(x, y) = x^4 + x^2y + y^2$.

Then $f_x = 4x^3 + 2xy$ and $f_y = x^2 + 2y$. Hence $f_x = 0 = f_y$ when $x = 0 = y$. At (0, 0), $f_{xx} = 0, f_{yy} = 2, f_{xy} = 0$ and $f_{xx}f_{yy} - f_{xy}^2 = 0$. This is the doubtful case and requires a different treatment.

We can write $f(x, y) = (x^2 + \frac{1}{2}y)^2 + \frac{3}{4}y^2 \geqslant 0$. Therefore f has a minimum value of 0 at (0, 0).

The above method for finding extreme values of a function of two variables can be extended to a function of n variables. A function $f(x_1, x_2, \ldots, x_n)$ of n variables has a stationary value at points making $\mathrm{d}f = 0$. That is, where

$$\frac{\partial f}{\partial x_1}\mathrm{d}x_1 + \frac{\partial f}{\partial x_2}\mathrm{d}x_2 + \cdots + \frac{\partial f}{\partial x_n}\mathrm{d}x_n = 0.$$

If $x_1, x_2, \ldots, x_n$ are independent variables, this implies that

$$\frac{\partial f}{\partial x_1} = \frac{\partial f}{\partial x_2} = \cdots = \frac{\partial f}{\partial x_n} = 0$$

is a necessary condition for an extreme value of f.

Suppose these derivatives vanish at $(a_1, a_2, \ldots, a_n)$. Then by Taylor's theorem,

$$\Delta f = f(a_1+h_1, a_2+h_2, \ldots, a_n+h_n) - f(a_1, a_2, \ldots, a_n)$$
$$= \frac{1}{2}\sum_{i=1}^{n}\left(h_i \frac{\partial^2 f}{\partial x_i^2} + 2h_ih_j \frac{\partial^2 f}{\partial x_i\,\partial x_j}\right) + \text{terms of the order } \rho^3,$$

where the second derivatives are all evaluated at $(a_1, a_2, \ldots, a_n)$, and $\rho = |h_1| + |h_2| + \cdots + |h_n|$. Then, in a sufficiently small neighbourhood of $(a_1, a_2, \ldots, a_n)$ the sign of Δf is the same as that of the quadratic form

$$E = \sum \left(h_i \frac{\partial^2 f}{\partial x_i^2} + 2h_ih_j \frac{\partial^2 f}{\partial x_i\,\partial x_j}\right) \equiv \sum h_ih_jf_{ij},$$

provided that E does not vanish. If E is positive definite, the extreme value is a minimum. This is so if

$$f_{11}, \quad \begin{vmatrix} f_{11} & f_{12} \\ f_{21} & f_{22} \end{vmatrix}, \quad \begin{vmatrix} f_{11} & f_{12} & f_{13} \\ f_{21} & f_{22} & f_{23} \\ f_{31} & f_{32} & f_{33} \end{vmatrix}, \ldots, \begin{vmatrix} f_{11} & f_{12} & \cdots & f_{1n} \\ \vdots & \vdots & & \vdots \\ f_{n1} & f_{n2} & \cdots & f_{nn} \end{vmatrix}$$

are all positive. (See, for example, Ferrar (1941)).

Example 3 Show that $f(x, y, z) = x^2 + y^2 + z^2 - 2xyz$ has a minimum at $(0, 0, 0)$.

$$f_x = 2(x - yz), \quad f_y = 2(y - zx), \quad f_z = 2(z - xy)$$
$$f_{xx} = f_{yy} = f_{zz} = 2, \quad f_{xy} = -2z, \quad f_{yz} = -2x, f_{zx} = -2y.$$

Then at $(0, 0, 0)$, $f_x = f_y = f_z = 0$;

$$f_{xx} = 2, \quad \begin{vmatrix} f_{xx} & f_{xy} \\ f_{yx} & f_{yy} \end{vmatrix} = 4, \quad \begin{vmatrix} f_{xx} & f_{xy} & f_{xz} \\ f_{yx} & f_{yy} & f_{yz} \\ f_{zx} & f_{zy} & f_{zz} \end{vmatrix} = 8.$$

As these are all positive f has a minimum at the origin.

10.12 Conditional maxima and minima

A more complicated problem arises when we wish to determine the extreme values of a function $f(x_1, x_2, \ldots, x_n)$ of n variables, the variables being connected by m $(<n)$ relations

$$\phi_r(x_1, x_2, \ldots, x_n) = 0, \qquad r = 1, 2, \ldots, m.$$

At a stationary value $df = 0$, that is

$$\frac{\partial f}{\partial x_1} dx_1 + \frac{\partial f}{\partial x_2} dx_2 + \cdots + \frac{\partial f}{\partial x_n} dx_n = 0. \tag{1}$$

'rom the equations of condition

$$\frac{\partial \phi_r}{\partial x_1} dx_1 + \frac{\partial \phi_r}{\partial x_2} dx_2 + \cdots + \frac{\partial \phi_r}{\partial x_n} dx_n = 0. \qquad r = 1, 2, \ldots, m. \tag{2}$$

/ultiply eqs. (2) by constants λ_r $(r = 1, 2, \ldots, m)$ and add to eqs. (1) to ;ive

$$\sum_{s=1}^{n} \left(\frac{\partial f}{\partial x_s} + \sum_{r=1}^{m} \lambda_r \frac{\partial \phi_r}{\partial x_s} \right) dx_s = 0.$$

Jow choose the constants λ_r so as to make the coefficients of dx_s $s = 1, 2, \ldots, m)$ zero. That is

$$\frac{\partial f}{\partial x_s} + \sum_{r=1}^{m} \lambda_r \frac{\partial \phi_r}{\partial x_s} = 0, \qquad s = 1, 2, \ldots, m$$

ınd

$$\sum_{s=m+1}^{n} \left(\frac{\partial f}{\partial x_s} + \sum_{r=1}^{m} \lambda_r \frac{\partial \phi_r}{\partial x_s} \right) dx_s = 0. \tag{3}$$

Since the n variables $x_1, x_2, \ldots, x_n$ are subject to m conditions, $n - m$ of the variables are independent; say $x_{m+1}, x_{m+2} \ldots, x_n$. Therefore the coefficients of these differentials in eq. (3) must all vanish, so that

$$\frac{\partial f}{\partial x_s} + \sum_{r=1}^{m} \lambda_r \frac{\partial \phi_r}{\partial x_s} = 0, \qquad s = 1, 2, \ldots, n.$$

These n equations together with the m equations of condition,

$$\phi_r(x_1, x_2, \ldots, x_n) = 0, \qquad r = 1, 2, \ldots, m,$$

are sufficient, in general, to determine the $m + n$ unknowns

$$x_1, x_2, \ldots, x_n, \lambda_1, \lambda_2, \ldots, \lambda_m.$$

This method is known as *Lagrange's method of undetermined multipliers.*

Example Find the points on $x^2 + y^2 + z^2 = 1$ at which the function xyz is stationary.

Let $$f = xyz, \quad \phi \equiv x^2 + y^2 + z^2 - 1 = 0$$

and consider $E = f + \lambda\phi$. The stationary values of f are the roots of the four equations

$$x^2+y^2+z^2 = 1 \tag{1}$$

$$E_x = yz+2\lambda x = 0 \tag{2}$$

$$E_y = zx+2\lambda y = 0 \tag{3}$$

$$E_z = xy+2\lambda z = 0. \tag{4}$$

Multiply (2), (3), (4) by x, y, z respectively and add to give

$$3xyz+2\lambda(x^2+y^2+z^2) = 0.$$

Using (1) this reduces to

$$2\lambda+3xyz = 0; \tag{5}$$

thus the required stationary values of $f = -\frac{2}{3}\lambda$.

Now (2), (3), (4) with (5) can be written,

$$yz(1-3x^2) = 0, \quad zx(1-3y^2) = 0, \quad xy(1-3z^2) = 0,$$

which yield the 8 stationary points $x = \pm 1/\sqrt{3}, y = \pm 1/\sqrt{3}, z = \pm 1/\sqrt{3}$, together with the 6 stationary points when two of the three variables x, y, z vanish, viz. $(\pm 1, 0, 0)$, $(0, \pm 1, 0)$, $(0, 0, \pm 1)$.

To illustrate how to determine the nature of these stationary values we consider the two points $(1/\sqrt{3}, 1/\sqrt{3}, 1/\sqrt{3})$ and $(1, 0, 0)$.

$$\begin{aligned}\Delta E &= E(x+h_1, y+h_2, z+h_3)-E(x,y,z)\\ &= \tfrac{1}{2}[E_{xx}h_1^2+E_{yy}h_2^2+E_{zz}h_3^2+2(E_{xy}h_1h_2+E_{yz}h_2h_3+E_{zx}h_3h_1)]\\ &\quad + \text{higher powers of } h_1, h_2, h_3.\end{aligned}$$

We must examine the sign of ΔE in a small neighbourhood of the stationary point when h_1, h_2, h_3 are subject to $d\phi = \phi_x h_1+\phi_y h_2+\phi_z h_3 = 0$, that is $xh_1+yh_2+zh_3 = 0$.

From (2), (3), (4) we have

$$E_{xx} = E_{yy} = E_{zz} = 2\lambda, \text{ and}$$

$$E_{xy} = z, \quad E_{yz} = x, \quad E_{zx} = y.$$

At $(1/\sqrt{3}, 1/\sqrt{3}, 1/\sqrt{3})$ we have $\lambda = -1/2\sqrt{3}$ (from (5)) and

$$\Delta E = -\frac{1}{\sqrt{3}}(h_1^2+h_2^2+h_3^2)+\frac{2}{\sqrt{3}}(h_1h_2+h_2h_3+h_3h_1)+\cdots,$$

subject to $h_1+h_2+h_3 = 0$. Now

$$\begin{aligned}2(h_1h_2+h_2h_3+h_3h_1) &= (h_1+h_2+h_3)^2-(h_1^2+h_2^2+h_3^2),\\ &= -(h_1^2+h_2^2+h_3^2) \qquad \text{since } h_1+h_2+h_3 = 0\end{aligned}$$

lence

$$\Delta E = -\frac{2}{\sqrt{3}}(h_1^2 + h_2^2 + h_3^2) + \text{higher powers of } h_1, h_2, h_3$$

$$< 0 \qquad \text{in a sufficiently small neighbourhood of } \left(\frac{1}{\sqrt{3}}, \frac{1}{\sqrt{3}}, \frac{1}{\sqrt{3}}\right).$$

hat is the stationary value at $(1/\sqrt{3}, 1/\sqrt{3}, 1/\sqrt{3})$ is a maximum. At ., 0, 0), $\lambda = 0$ and

$$\Delta E = h_2 h_3 + \text{higher powers of } h_1, h_2, h_3,$$

ıbject to $h_1 = 0$. Clearly $\Delta E \gtrless 0$ according as h_2 and h_3 have the same or pposite signs, and so the stationary value is neither a maximum nor a inimum.

XERCISES 10(d)

1. Find, and determine the nature of, the stationary points of the following functions:

   (a) $x^2 + y^2 + 2x + 4y + 10$ (b) $x^2 + 3xy + y^2$

   (c) $x^3 + y^3 + 3xy$ (d) $x^4 - 3x^2y + y^3$

   (e) $x^4 + y^4 - 6(x^2 + y^2) + 8xy$ (f) $x^2 + y^2 + x^2y^2 - 2xy \operatorname{ch}^2 \alpha$ $(\alpha \neq 0)$

   (g) $\sin^2 x \cos y + \sin^2 y \cos x$, $0 \leqslant x < \pi, 0 \leqslant y < \pi$

2. Show that $x^4 + y^4 + z^4 - 4xyz$ has a minimum at $(1, 1, 1)$.

3. Find the range of values of the constant λ for which the function $\lambda(e^y - 1) \sin x - \cos x \cos 2y + e^{z^2}$ has a minimum at the origin.

4. Find the stationary values of $(x-1)^2 + (y-2)^2 + (z-2)^2$ when $x^2 + y^2 + z^2 = 9$.

5. Given $f(x, y) = (y^2 - 2ax)(y^2 - 2bx), 0 < a < b$, show that f has a minimum value at the origin if $y = mx$, where m is a constant. Show also that the origin is not a minimum when unrestricted variation of x and y is allowed.

6. Show that $V = 2xy + yz + zx$ has a maximum at $(\frac{1}{2}, \frac{1}{2}, 0)$ when x, y, z satisfy the condition $x + y + z = 1$.

7. Show that on $x^3 + y^3 + z^3 - 2xyz = a^3$ $(a > 0)$ the function $x + y + z$ is stationary at (a, a, a) and three other points. Prove that the function is a maximum at (a, a, a).

8. Show that the function $x^2 + y^2 + z^2$ has stationary values at $x = y = -z = \pm 1$ when x, y, z are subject to the two conditions $xy = 1$ and $z(x + y) + 2 = 0$. Show that these stationary values are minima.

Bibliography and Suggestions for Further Reading

Brown, A.L. and Page, A. *Elements of Functional Analysis*, Van Nostrand Reinhold, 1970.

Copson, E.T., *Metric Spaces*, Cambridge University Press, 1968.

Dieudonné, J., *Foundations of Modern Analysis*, Academic Press, 1960.

Ferrar, W.L., *Algebra*, Oxford University Press, 1941.

Gelbaum, B.R. and Olmsted, J.M.H., *Counterexamples in Analysis*, Holden-Day, 1964.

Hardy, G.H., *Pure Mathematics*, Cambridge University Press (10th ed.), 1952.

Hatton, M.D., *Elementary Mathematics for Scientists and Engineers*, Pergamon Press, 1965.

Isaacs, G.L., *Real Numbers*, McGraw-Hill, 1968.

Landau, G.H., *Foundations of Analysis*, Chelsea, 1951.

MacDuffee, C.C., *An Introduction to Abstract Algebra*, Wiley, 1948.

Pitts, C.G.C., *Introduction to Metric Spaces*, Oliver and Boyd, 1972.

Rudin, W., *Principles of Mathematical Analysis*, McGraw-Hill, 1964.

Russell, B., *Introduction to Mathematical Philosophy*, Allen and Unwin, 1920.

Simmons, G.F., *Introduction to Topology and Modern Analysis*, McGraw-Hill, 1963.

Thurston, H.A., *The Number System*, Dover, 1967.

Answers and Hints for Solutions

Exercises 1(a)

1. (a) Any integer can be written in only one of the forms $3k$, $3k-1$, $3k-2$, where k is an integer. Suppose $r = p/q$; taking $p = 3k-1$ leads to $3m = 3n-1$ which is impossible for integers m and n.

2. $2{\cdot}5 \times 10^{-1}$, $1{\cdot}64 \times 10^{-2}$, $2{\cdot}225 \times 10^{-3}$, $2{\cdot}445 \times 10^{-4}$, $1{\cdot}821 \times 10^{-5}$, $1{\cdot}238 \times 10^{-6}$

3. Let $q = p - h$, where $0 < h < 1$; then
$q^2 = p^2 - 2ph + h^2 > p^2 - (p^2 - 2) + h^2 > 2$ if $h < \frac{1}{2}(p^2 - 2)/p$.

4. $q - p = 2p(2 - p^2)/(3p^2 + 2)$, $q^2 - 2 = (p^2 - 2)^3/(3p^2 + 2)^2$

5. If $b \leqslant a$, Section 1.3(b) shows b is in α

6. Since $\alpha < \beta$, there is a rational b in β which is not in α. Similarly there is a rational c in γ which is not in β. Hence $b < c$ and so c is not in α.

7. Let c be in $\alpha + 0$. Then $c = a + b$, where a is in α and b is in 0. Then $b < 0$ and so $a + b < a$; that is c is in α.

Exercises 1(b)

9. $2^{n+1} - n - 2$ 10. 'Proof' requires $n \geqslant 2$

Exercises 1(c)

1. $x < 3$ 2. $|x| \leqslant 2$ 3. $-2 < x < 3$ 4. All x 5. No x

6. $x<1$ or $\frac{3}{2}<x<2$ 7. $x>\frac{1}{2}$ or $x<-5$ 8. $-5<x<1$

9. $1<x<7, x\neq 4$ 10. $x<-\frac{5}{3}$ or $x>5$ 11. $x>-\frac{1}{2}$

12. $-\frac{2}{3}<x<5$ 13. $-1<x<\frac{5}{3}$ or $x<-3$ 14. $6<x<8$

18. (b) $\dfrac{1}{1-x}<1+2x$ implies $0<x<\frac{1}{2}$

19. Use $(a^2-b^2)^2\geqslant 0$. Use Cauchy's inequality

20. (b) $(1-a_1)(1-a_2)>1-(a_1+a_2)$; use induction

21. $1-a_r^2<1$ and $a_r<1$, hence $1\pm a_r<1/(1\mp a_r)$.

Exercises 2(a)

1. 1, 10^3, 10^6; 10^6, 10^6, 10^7; 10^{12}, 10^9, 10^8; 10^{18}, 10^{12}, 10^9.

6. (a) $f(x)=x, 0<x<1$; $f(x+1)=f(x)$ for all x.
(b) $xH(x)-H(x-1)-H(x-2)-\cdots$.

7. Let $P(0)=k$; then since $P(x+a)=P(x)$ for all x, $P(x)=k$ has infinitely many roots $x=0, a, 2a, 3a, \ldots$. But a polynomial of degree n cannot have more than n zeros unless it is identically zero.

8. (b) and (c) Use $\operatorname{ch} x\pm\operatorname{sh} x=e^{\pm x}$.

10. (a) inc. $m>0$, dec. $m<0$ (b) inc. (c) dec. (d) dec. (e) inc. (f) dec.

Exercises 2(b)

1. (a) $-\frac{1}{2}, -1$ (b) $1, \frac{1}{4}$ (c) f unbounded (d) $2/\pi$, (e) 1, 0.

2. If $f>0(<0)$ for all x, $|f|=f(-f)$ and equality holds. If f takes positive and negative values, $\inf f<0$, $\inf|f|\geqslant 0$, $\sup|f|=\sup f$ or $-\inf f$.

3. (a) Let $\inf f=m$, $\sup f=M$, $\inf g=n$, $\sup g=N$, then $f+g\leqslant M+N$. (b) Let $\sup(f+g)=\Lambda$, then $\Lambda\leqslant M+N$. Also $f+g<(m+\epsilon)+N$ for at least one x; since ϵ is arbitrary, $f+g\leqslant m+N$.

4. Let $y=0$; $f(x)=f(x)f(0)$ for all x. Hence $f(0)=1$. Put $x=y=\frac{1}{2}a$, $f(a)=[f(\frac{1}{2}a)]^2\geqslant 0$. If $f(a)=0$, then $f(x)\equiv 0$ (put $x=a, y=x-a$).

Suppose $f < M$ for all x, then $f(-x) = [f(x)]^{-1} > M$, a contradiction unless $f \equiv 1$.

Exercises 2(c)

1. (a) $(x^3+7)^{-1} < x^{-3} < \epsilon$ when $x > X = \epsilon^{-1/3}$

(b) $x^2 - 2x - 1 > x^2 - 3x \ (x > 1) > \frac{1}{2}x^2 \ (\frac{1}{2}x^2 > 3x)$.
Hence $(x^2-2x-1)^{-1} < 2x^{-2} < \epsilon$ when $x > X = \max(6, \sqrt{(2/\epsilon)})$

(c) $\dfrac{4x^3-1}{3x^5+5} < \dfrac{4x^3}{3x^5} < \epsilon$ when $x > X = \sqrt{(4/3\epsilon)}$

(d) $|f(x) - l| < \dfrac{3x^3+5}{4x^4-x} < \dfrac{4x^3}{3x^4} < \epsilon$ when $x > X = \max(2, 4/3\epsilon)$

(e) $|f(x) - l| < \dfrac{7x}{4x^3} \ (x > 11) < \epsilon$ when $x > X = \max(11, \sqrt{(7/4\epsilon)})$

(f) $|f(x) - l| < \dfrac{6x}{\frac{1}{2}x^2} < \epsilon$ when $x > X = \max(\sqrt{2}, 12/\epsilon)$

(g) $|f(x) - l| < \dfrac{6x}{|x^2-6x|} \ (x > 6) < \dfrac{6x}{\frac{1}{2}x^2} \ (x > 12) < \epsilon$
when $x > X = \max(12, 12/\epsilon)$

(h) $|f(x) - l| < \dfrac{2x^2+10x}{3|3x^3-2x^2|} \ (x > 10) < \dfrac{3x^2}{6x^3} < \epsilon$
when $x > X = \max(10, 1/2\epsilon)$

(i) $\left|\dfrac{[x]-x}{x}\right| < \dfrac{1}{x} < \epsilon$ when $x > X = 1/\epsilon$

(j) $\left|\dfrac{\sin x}{x}\right| \leqslant \dfrac{1}{|x|}$ for all x

(k) Put $x = -y$

(l) See (j)

(m) $f(x) > 3x^5/4x^3 > K$ when $x > X = \sqrt{(4K/3)}$

(n) Put $x = -y$, $-\dfrac{y^4}{2y^3-1} < -K$ when $y > X = 2K$;
thus $x^4/(4x^3-1) < -K$ when $x < -X = -2K$.

2. $$|a_1x^{m-1} + a_2x^{m-2} + \cdots + a_m| \leqslant x^{m-1}(|a_1| + \cdots + |a_m|) \ (x > 1) \leqslant Ax^{m-1} < \tfrac{1}{2}x^m$$
when $x > X = \max(1, 2A)$

3. Use result of No. 2. When $m = n$, limit is a_0/b_0.

5. $f(x) > K$ for $x > X$; hence $1/f(x) < 1/K = \epsilon$ when $x > X$.
As $x \to \infty$ $(\sin x)/x \to 0$, but $x \operatorname{cosec} x$ oscillates.

Exercises 2(d)

1. (a) $|(x^2-2x-1)+1| = |x|.|x-2| < 3|x-2|$ $(|x-2|<1) < \epsilon$ when $|x-2| < \delta = \min(1, \frac{1}{3}\epsilon)$
(b) $|(x^3-3x^2+5x+4)-7| \leqslant |x-1|(|x-1|^2+2) < 3|x-1|$ $(|x-1|<1) < \epsilon$ when $|x-1| < \delta = \min(1, \frac{1}{3}\epsilon)$.
(c) $|(x^2+3x+4)-2| \leqslant |x+2|(|x+2|+1) < \epsilon$ when $|x+2| < \delta = \min(1, \frac{1}{2}\epsilon)$.
(d) $|x \sin(1/x)| \leqslant |x|$.
(e) $\left|\dfrac{4x^2-1}{3x^2+5}+\dfrac{1}{5}\right| < \dfrac{23x^2}{5.5}$ $(x^2>0) < \epsilon$ when $|x| < \delta = 5\sqrt{(\epsilon/23)}$.
(f) $\left|\dfrac{3x^3-3x+4}{2x^3+1}-\dfrac{4}{3}\right| = \left|\dfrac{(x-1)^3+3(x-1)^2-6(x-1)}{3(2x^3+1)}\right| \leqslant \dfrac{10|x-1|}{3}$ $(|x-1|<1) < \epsilon$ when $|x-1| < \delta = \min(1, 3\epsilon/10)$.
(g) $\left|\dfrac{x^2-3x+2}{x^3-3x^2+2}-\dfrac{1}{3}\right| \leqslant \dfrac{|x-1|^2(|x-1|+3)}{3|x-1||x^2-2x-2|} < \dfrac{4|x-1|}{6}$ $(|x-1|<1) < \epsilon$ when $|x-1| < \delta = \min(1, 3\epsilon/2)$.
(h) $\left|\dfrac{x^2+5x+4}{x^2+3x+2}\right| < \dfrac{4|x+4|}{2}$ $(|x+4|<1) < \epsilon$ when $|x+4| < \delta_1 = \min(1, \frac{1}{2}\epsilon)$.

2. If $\lim f = l_1$ and l_2, $l_1 \neq l_2$, $|f(x)-l_1| < \frac{1}{2}\epsilon$ when $|x-a| < \delta_1$ $|f(x)-l_2| < \frac{1}{2}\epsilon$ when $|x-a| < \delta_2$. Then, if $\delta = \min(\delta_1, \delta_2)$ $|l_1-l_2| \leqslant |f(x)-l_1| + |f(x)-l_2| < \epsilon$ when $|x-a| < \delta$. Since ϵ is arbitrary this implies $l_1 = l_2$.

3. Take $\epsilon = \frac{1}{2}l$ in the definition of limit, Section 2.6.

4. Suppose $l < 0$ and take $\epsilon = -\frac{1}{2}l$ in the limit definition.

5. $l-\epsilon < f < l+\epsilon$, $|x-a| < \delta_1$ and $L-\epsilon < g < L+\epsilon$, $|x-a| < \delta_2$. If $l > L$, then $0 < l-L < (f+\epsilon)-(g-\epsilon)$ and so $f-g > -2\epsilon \geqslant 0$, a contradiction.

6. $-\epsilon < f-l < \epsilon$, $|x-a| < \delta_1$, $-\epsilon < h-l < \epsilon$, $|x-a| < \delta_2$. Then $-\epsilon < f-l \leqslant g-l \leqslant h-l < \epsilon$ for $|x-a| < \delta = \min(\delta_1, \delta_2)$.

7. $k > 0$, $|f-l| < \epsilon/k$ for $|x-a| < \delta$.

8. Use $||f|-|l|| \leqslant |f-l|$. Converse is false, e.g. $f(x) = -1$, x rational, $f(x) = 1$, x irrational.

9. $\sup f = M$ then $f \leqslant M$ for all x and $f > M-\epsilon$ for some x_1. But f is increasing, so $f > M-\epsilon$ for $x > x_1$. Thus $M-\epsilon < f \leqslant M$ for $x > x_1$.

Exercises 2(e)

1. Let $h > 0$. (a) $f(3-h) = (3-h-[3-h]) = 1-h \to 1$ as $h \to 0$.
(b) $f(1+h) = ([1+h]+1)^{-1} = \frac{1}{2}$. (c) $f(1-h) = -1$. (*d*) $-\infty$
(e) $-\infty$ (f) 2 (g) 1
(h) $f(2-h) = 2-h+(2-h-[2-h])^2 = 3-3h+h^2 \to 3$ as $h \to 0$.

2. $f(0-) = 0, f(0+) = -4; f(1-) = 1 = f(1+)$;
$f(2-) = 1, f(2+) = 2; f(3-) = 2, f(3+) = 3$;
$f(4-) = 3 = f(4+); f(5-) = 4 = f(5+)$.

3. See Section 2.8.

4. (a) -1 (b) 0, put $x = t^{-1}$ and let $t \to 0$ (c) $\frac{1}{3}$ (d) 2 (e) 5/2
(f) 0, since th $x \to 1$ as $x \to \infty$.

5. If n is a positive integer,
$x^n - a^n = (x-a)(x^{n-1} + ax^{n-2} + a^2x^{n-3} + \cdots + a^{n-1})$.
If $n = -m$, m a positive integer,
$x^{-m} - a^{-m} = -(x^m - a^m)/(a^m x^m)$.
If $n = p/q$, let $x = y^q$, $a = b^q$; then

$$\frac{x^n - a^n}{x-a} = \left(\frac{y^p - b^p}{y-b}\right) \Bigg/ \left(\frac{y^q - b^q}{y-b}\right).$$

Exercises 3(a)

1. $|x^2 - 4| = |x+2|\,|x-2| < 5|x-2|$ $(|x-2| < 1) < \epsilon$
for $|x-2| < \delta = \min(1, \frac{1}{5}\epsilon)$.

2. $$\frac{|1+x|\,|1-x|}{x^2} < \frac{5|x-1|}{\frac{1}{4}} \quad (|x-1| < \tfrac{1}{2}) < \epsilon$$
when $|x-1| < \delta = \min(\frac{1}{2}, \epsilon/20)$.

3. $|\,|x| - |-1|\,| \leqslant |x+1|$.

4. $$|\sqrt{x} - \sqrt{2}| = \frac{x-2}{\sqrt{x}+\sqrt{2}} < \frac{|x-2|}{1+\sqrt{2}} \quad (x > 1) < \epsilon$$
when $|x-2| < \delta = \min(1+\sqrt{2})\epsilon)$.

5. $|x^4 - a^4| = |x-a|\,|x+a|\,|x^2+a^2| < \frac{5}{2} \cdot \frac{7}{4}|a|^3|x-a|$ $(|x-a| < \frac{1}{2}|a|$,
i.e. $\frac{1}{2}|a| < |x| < \frac{3}{2}|a|)$
$< \epsilon$ when $|x-a| < \delta = \min(\frac{1}{2}|a|, 8\epsilon/35|a|^3)$.
If $a = 0$, $|x^4 - a^4| = |x|^4 < \epsilon$ when $|x| < \delta = \epsilon^{1/4}$.

6. $\left|\dfrac{x^2-4}{x-1}-4\right| = |x|\left|\dfrac{x-4}{x-1}\right| < 7|x| \; (|x|<\tfrac{1}{2}) < \epsilon$

when $|x| < \delta = \min(\tfrac{1}{2}, \tfrac{1}{7}\epsilon)$

$\left|\dfrac{x^2-4}{x-1}\right| = |x-2| . \left|\dfrac{x+2}{x-1}\right| < 9|x-2| \; (|x-2|<\tfrac{1}{2}) < \epsilon$

when $|x-2| < \delta = \min(\tfrac{1}{2}, \tfrac{1}{9}\epsilon)$

7. All x except $x = 2$, where the function is undefined.

8. Discontinuous at each integer. 9. $a \leqslant x \leqslant b$ 10. $a \leqslant x < b$

11. See Ex. 2(d) No. 3. 12. See Section 2.6.

13. Let $0 \leqslant a \leqslant 1, a \neq \tfrac{1}{2}$. There are rationals and irrationals arbitrarily close to a. So f takes values arbitrarily close to both a and $1-a$; these values differ by $|1-2a|$ which is zero only when $a = \tfrac{1}{2}$.
$|f(\tfrac{1}{2}+h) - f(\tfrac{1}{2})| = |h|$ whether h is rational or irrational.

14. Let $\epsilon > 0$; then if a is any irrational $|f(x) - f(a)| = 0$ for all irrational x. The set of integers q such that $1/q \geqslant \epsilon$ is finite, $N = [\epsilon^{-1}] + 1$; these values of q correspond to a finite set of values of p/q, since $p/q \leqslant 1$. Let δ be distance from a to the nearest of these points. Then for rational $x = p/q$, $|f(x) - f(a)| = 1/q < \epsilon$ when $|x - a| < \delta$.

Exercises 3(b)

1. and 2. See Section 2.8.

4. $f(x) = \log(x+2), 0 \leqslant x < 1; f(x) = -\sin x, 1 < x \leqslant \tfrac{1}{2}\pi;$
$f(1) = \tfrac{1}{2}[\log 3 - \sin 1]$.

5. Let ϕ be either $\max(f, g)$ or $\min(f, g)$; then
$|\phi(x) - \phi(a)| \leqslant |f(x) - f(a)| + |g(x) - g(a)|$.

6. Let $a < \xi < b$; then since f is increasing the one-sided limits exist and $f(\xi + 0) \geqslant f(\xi)$; also $f(x) \geqslant f(\xi)$ when $x > \xi$. If $f(\xi + 0) = f(\xi) + k$, $k > 0$, then since f is increasing, $f(x) \leqslant f(\xi)$ when $x < \xi$ and so f does not take values between $f(\xi)$ and $f(\xi) + k$, contrary to hypothesis.

Exercises 3(c)

1. $(x-c)/m$. 2. x^2. 3. $1/x$. 4. $(2x+1)/(x-1), x>1$.

5. $\frac{1}{2}[1-\sqrt{(1-2x^2)}]\,x^{-1}, |x| \leqslant \frac{1}{2}$.

6. $|f(x)-f(x')| = |x-x'|\,|x+x'| \leqslant 4|x-x'| < \epsilon$ for $|x-x'| < \delta = \frac{1}{4}\epsilon$, since $\sup|x+x'| = 4$ in $0 \leqslant x \leqslant 2$.

7. $|f(x)-f(x')| = \dfrac{|x-x'|}{\sqrt{x}+\sqrt{x'}} \leqslant \frac{1}{2}|x-x'|$ for $1 \leqslant x \leqslant 2$.

8. Use $\sqrt{|x-x'|} \leqslant \sqrt{x}+\sqrt{x'}$.

9. $|f(x)-f(x')| \leqslant |x-x'|$ for $x \geqslant 1$.

10. See No. 6; here $\sup|x+x'| = \infty$ when x, x' are in the unbounded interval $x \geqslant 1$.

11. $|\sin(k-\frac{1}{2})\pi - \sin 3(k-\frac{1}{2})\pi| = 2$ and $|x-x'| < \delta$ when integer $2k-1 > \dfrac{4}{3\pi\delta}$.

Exercises 4(a)

1. $4x^3$ (See Exercise 2(e) No. 5). 2. $-\frac{1}{2}x^{-3/2}$ (See Example 3, Section 4.1
3. $x^{1/2}$. 4. $2\cos 2x$ (using $\lim\limits_{\theta\to 0}(\sin\theta)/\theta = 1$)

5. $\dfrac{\tan(x+h)-\tan x}{h} = \dfrac{\tan h}{h}\{1+\tan x\tan(x+h)\} \to \sec^2 x$ as $h \to 0$.

6. $\dfrac{\operatorname{ch}(x+h)-\operatorname{ch}x}{h} = \operatorname{sh}(x+\frac{1}{2}h)\dfrac{\operatorname{sh}\frac{1}{2}h}{\frac{1}{2}h} \to \operatorname{sh}x$ as $h \to 0$.

7. $\dfrac{f(h)-f(0)}{h} = h^{-2/3}$ which has no finite limit as $h \to 0$.

8. See Example 3, Section 3.1; $h^{-1}\{f(h)-f(0)\} = \sin(1/h)$, which has no limit as $h \to 0$.

9. $f'(0) = \lim\limits_{h\to 0} h\sin(1/h) = 0$.

10. $f'_{-}(0) = \lim\limits_{h\to 0-}\tan^{-1}(1/h) = -\frac{1}{2}\pi; f'_{+}(0) = \frac{1}{2}\pi$.

11. $g'_-(2) = 1$; $g'_+(2)$ does not exist.

12. $h'_-(0) = 1$; $h'_+(0) = 0$.

Exercises 4(b)

1. (a) $(x+1)^{-1}(x^2-1)^{-1/2}$ (b) $4x(4-3x^2)(x^2-1)(2-x^2)^3$
(c) $\frac{1}{2}(5x^2+3)(x+1/x)^{-1/2}$ (d) $e^{-3x}(2\cos 2x - 3\sin 2x)$
(e) $\frac{1}{3}\sec^2\frac{1}{3}x\exp(\tan\frac{1}{3}x)$ (f) $2\cos x(1-\sin x)^{-2}$
(g) $\cot x$ (h) $\sin(\log x)+\cos(\log x)$ (i) $(x^2-1)^{-1}$ (j) $2x/(1+x^4)$
(k) $2x\sin^{-1}(2x)+2x^2(1-4x^2)^{-1/2}$ (l) $D\sin^{-1}(1/x) = -1/[x\sqrt{(x^2-1)}]$
(m) $D(\frac{1}{2}x+\frac{1}{4}\pi) = \frac{1}{2}$ (n) $(x^{-1}\sin x + \cos x\log x)x^{\sin x}$
(o) $D\tan^{-1}(x^2/a^2) = 2a^2x/(a^4+x^4)$.

2. (a) $\mathrm{ch}x + x\mathrm{sh}x$ (b) $2x\,\mathrm{sh}\,x + x^2\,\mathrm{ch}\,x$ (c) $\frac{1}{2}x^{1/2}\,\mathrm{sec}h^2\frac{1}{2}x + \frac{1}{2}x^{-1/2}\,\mathrm{th}\frac{1}{2}x$
(d) $(\mathrm{sh}\,x+1)/[\mathrm{ch}\,x(\mathrm{sh}\,x-1)]$ (e) $\frac{1}{2}(e^x+5e^{-5x})$ (f) $\frac{1}{2}x^{-1/2}(x-1)^{-1/2}$
(g) $\mathrm{th}^{-1}x$ (h) $2x+2(2x^2+1)(x^2+1)^{-1/2}\mathrm{sh}^{-1}x$ (i) $(1-e^{-2x})^{-1/2}$
(j) $\frac{1}{2}\sec x$ (k) $\sec x$ (l) $\dfrac{2}{(1-x)^2}\sec\left(\dfrac{1+x}{1-x}\right)$.

5. $-a^{-1}\,\mathrm{cosec}^3\,t$.

6.(a) $(x^2-8x+12)e^{-x}$. (b) $16(2x^3+15x^2+30x+15)e^{2x}$
(c) $(x^2-12)\cos x + 8x\sin x$ (d) $12x^{-3}$
(e) $\left[\dfrac{12}{1+x^2}+\dfrac{12x-2}{(1+x^2)^2}+\dfrac{8x^2}{(1+x^2)^3}-8\tan^{-1}x\right]e^{-2x}$
(f) $(-1)^n e^{-x}(x^2-2nx+n^2-n)$
(g) $2^{n-3}e^{2x}[8x^3+12nx^2+6n(n-1)x+n(n-1)(n-2)]$
(h) $[x^2-n(n-1)]\cos(x+\frac{1}{2}n\pi)+2nx\sin(x+\frac{1}{2}n\pi)$.

7. Use partial fractions.

8. Write $x^ne^{1/x} = xy$; prove by induction using the first part of the question.

9. Use induction.

10. Differentiate $(1+x^2)y' = 1$. Then
$y^{(2k)}(0) = 0, \quad y^{(2k-1)}(0) = (-1)^{k-1}(2k-2)!, \qquad k = 1, 2, 3, \ldots$.

11. Differentiate $(1-x^2)y'^2 = m^2(1-y^2)$;
$(1-x^2)y^{(n+2)} - (2n+1)xy^{(n+1)} - (n^2-m^2)y^{(n)} = 0$.
$y^{(2k-1)}(0) = 0, \quad y^{(2k)}(0) = -m^2(2^2-m^2)(4^2-m^2)\ldots(4k^2-m^2)$,
$k = 1, 2, 3, \ldots$.

Exercises 4(c)

1. Minimum -14 at $x = 0$.

2. Maximum 7 at $x = 0$; minimum -9 at $x = \pm 1$.

3. Minimum 12 at $x = 2$.

4. Minimum 8 at $x = 0$; maximum $60e^{-2}$ at $x = 2$.

6. $y' = -re^{-ax} \sin (bx + \alpha)$ where $r^2 = a^2 + b^2$, $\alpha = \tan^{-1} (b/a)$;
Extreme values at $x_n = (\alpha + n\pi)/b, n = 0, 1, 2, \ldots$.

7. Maximum at $x = 0$ when $0 \leqslant a \leqslant \frac{1}{2}$.
$a < 0$, maximum at $x = 0$, minima at $x = \pm\sqrt{(2 - 1/a)}$;
$a > \frac{1}{2}$, minimum at $x = 0$, maxima at $x = \pm\sqrt{(2 - 1/a)}$.

9. Consider the sign of the incremental ratio.

10. Use No. 9.

Exercises 4(d)

1. $f' = 0$ when $x = (m - n)/(m + n)$ which lies in $(-1, 1)$.

2. (a) f discontinuous at $x = 0$ (b) $g'(0)$ does not exist.

3. Suppose $u = 0$ at x_1, x_2 where $a \leqslant x_1 < x_2 \leqslant b$ and that v does not vanish in $[x_1, x_2]$. Apply Rolle's theorem to $f(x) = u/v$.

4. Any point of $[a, b]$; $\frac{1}{2}(a + b)$; $\sqrt{[\frac{1}{3}(a^2 + ab + b^2)]}$.

5. $1 - \frac{1}{6}\sqrt{21}$.

6. Consider $g(x) = f(x) - \gamma x$.

8. Use $f'''(x) = x^2 e^x$. Write $f(x) = (e^x - 1)(x^2 - 6x + 12) - 12x$, then $f > 0$ for $x > 0$.

10. Assume true for $1, 2, \ldots, n$. Then for n even, $P'_{n+1} = P_n \neq 0$, and so P_{n+1} is monotonic. Since $n + 1$ is odd, P_{n+1} changes sign, hence $P_{n+1} = 0$ once only. If n is odd, $P'_{n+1} = P_n = 0$ for $x = \xi$;

$$P_{n+1}(\xi) = P_n(\xi) + \frac{\xi^{n+1}}{(n+1)!} > 0,$$

for $n + 1$ is even. Hence minimum $P_{n+1} > 0$.

11. Let $u = (x^2 - 1)^n$; use Rolle's theorem and induction.

12. Apply the mean value theorem twice to yield $f''(x+\theta y) > \theta O$ for all x, y; in particular for $y = 0$.

13. Apply mean value theorem to $g(x) = f(x) - f(x-h)$ in $(x, x+h)$.

Exercises 4(e)

1. $\xi = 28/9$.

2. Theorem does not apply since f and g have a common zero at $x = 0$; substitution into the formula leads to $\xi = -\frac{1}{2}$.

3. 2   4. $-\frac{1}{2}$ 5. $\frac{1}{2}$ 6. $2 \log 2$ 7. $\frac{1}{2}$ 8. 0   9. $1/(\pi \cos n\pi)$ 10. -2

11. $e^{1/2}$ 12. e^{-1} 13. 1   14. $\frac{1}{2}$ 15. $b - \frac{1}{2}c \sin nt + \frac{1}{2}nct \cos nt$

16. e^x 17. $\log x$ (See Exercise 5(a) No. 4)

18. $f'(0) = 0$ if $m > 1$; f' continuous when $m > n+1$ $(n > 0)$. $f(x) = x^2 \sin(1/x)$, $g(x) = x$.

19. Use Cauchy's mean value theorem with $g(x) = e^x$.

20. Use second mean value theorem (Exercise 4(d) No. 14) to numerator and denominator of L.H.S. in Cauchy's mean value theorem. Use mean value theorem on R.H.S.

Exercises 4(f)

1. Use second mean value theorem (Exercise 4(d) No. 14) and the continuity of f'',
$\frac{1}{2}[f(x+h) + f(x-h)] = f(x) + \frac{1}{4}h^2[f''(x+\theta_1 h) + f''(x-\theta_2 h)]$,
$= f(x) + \frac{1}{2}h^2[f''(x) + \epsilon]$, $0 < \theta_1, \theta_2 < 1$ where $\epsilon \to 0$ as $h \to 0$.

2. See Section 4.11.

3. $f(x) < \left(\dfrac{x_2 - x}{x_2 + x_1}\right) f(x_1) + \left(\dfrac{x - x_1}{x_2 - x_1}\right) f(x_2)$; let $x_2 \to x + 0$ and $x \to x_1 + 0$, then $f(x) \leqslant f(x+0)$ and $f(x_1 + 0) \leqslant f(x_1)$. Hence $f(x) = f(x+0)$. Similarly $f(x) = f(x-0)$.

Exercises 5(a)

1. Cf. Exercise 2(c) No. 1.

2. Sequences converge to 1, 2, 3, 3 respectively (cf. Exercises 2(c))

3. (a) $\frac{1}{2}, 0$ (b) $\frac{1}{2}, -\frac{1}{2}$ (c) $\frac{2}{3}, 0$ (d) $1, -1$ (e) $\frac{4}{3}, -1$ (f) $\frac{13}{12}, -\frac{7}{6}$.

4. $x = (1 + x_n)^n$. For $x > 1$ use Bernoulli's inequality (Example 7, Section 1.7) (cf. Example 1, Section 5.3).
For $0 < x < 1$, write $x = 1/\mu$, $\mu > 1$ and use limit theorems (Section 2.8).

5. Remember $\lim_{n \to \infty} x^{1/n} = 1, x > 0$. For last part write
$f(x) + f(y) = f(x) - f(y^{-1})$ and use limit theorems.

7. Cf. Exercise 2(d) No. 2.

Exercises 5(b)

1. $\frac{1}{2}$ (cf. Section 2.8).

3. $x_n - y_n \to \alpha - \beta$ as $n \to \infty$. If $\alpha > \beta$, $x_n - y_n > 0$ for $n \geqslant N$, contradicting $x_n < y_n$ for all n. For example $x_n = 1/n, y_n = 1 + 1/n$; $x_n = 1/n, y_n = 2/n$.

4. $(x_n), (y_n)$ both converge to l so $l - \epsilon < x_n < l + \epsilon$ for $n \geqslant N_1$, and $l - \epsilon < y_n < l + \epsilon$ for $n \geqslant N_2$. Then if $n \geqslant N = \max(N_1, N_2)$,
$l - \epsilon < x_n \leqslant a_n \leqslant y_n < l + \epsilon$.

5. $||x_n| - |\alpha|| \leqslant |x_n - \alpha|$. $x_n = (-1)^n(1 + 1/n)$, then $|x_n| \to 1$ as $n \to \infty$, but x_n oscillates.
If $\alpha = 0$, note $|x_n| = ||x_n||$.

6. (a) 1 since $n^{1/n} \to 1$ as $n \to \infty$.
(b) e since $\dfrac{n+1}{\sqrt[n]{n!}} = \sqrt[n]{(y_1 y_2 \cdots y_n)}$ where $y_r = \left(\dfrac{r+1}{r}\right)^r$
(Example 2, Section 5.4)

Exercises 5(c)

1. Consider $x_{n+1} - x_n$, except for (i).
(a) decreasing for $n \geqslant 3$ (b) increasing

) increasing (d) decreasing
) decreasing (f) not monotonic
) not monotonic (h) increasing
) decreasing for $n \geqslant 10$ (consider x_{n+1}/x_n).

2. $x_{n+1} \geqslant K^{n+1-N} x_N$, and $K^n \to \infty$ as $n \to \infty$ since $K > 1$.

3. $x_{n+1} \leqslant K^{n+1-N} x_N$, and $K^n \to 0$ as $n \to \infty$ since $0 < K < 1$.

4. $l - \epsilon < x_n/x_{n+1} < l + \epsilon$ for $n \geqslant N$.
a) Choose ϵ so that $l + \epsilon = K^{-1} < 1$ and use No. 2.
b) Choose ϵ so that $l - \epsilon = K^{-1} > 1$ and use No. 3.

5. (a) Use No. 3 (b) Use No. 4.

6. (a_n) is increasing if $a_{n-1} < a_n$; if

$$\left(1 + \frac{1}{n-1}\right)^{-1} < \left(\frac{1 + 1/n}{1 + 1/n - 1}\right)^n \quad \text{or} \quad 1 - \frac{1}{n} < \left(1 - \frac{1}{n^2}\right)^n, \text{ which is true}$$

by Bernoulli's inequality. Similarly (b_n) is decreasing.

$$b_n - a_n = \left(1 + \frac{1}{n}\right)^n \left[\left(1 - \frac{1}{n^2}\right)^{-n} - 1\right] = \left(1 + \frac{1}{n}\right)^n \left[\left(1 + \frac{1}{n^2 - 1}\right)^n - 1\right]$$

$$> 2\left(1 + \frac{n}{n^2 - 1} - 1\right) > 0.$$

$$\text{Also } 0 < b_n - a_n = \left(1 - \frac{1}{n}\right)^{-n} \left[1 - \left(1 - \frac{1}{n^2}\right)^n\right] < \frac{1}{n}\left(\frac{n}{n-1}\right)^n$$

$$= \frac{1}{n-1} a_{n-1} < \frac{4}{n-1}.$$

$$a_n < 1 + 1 + \frac{1}{2!} + \cdots + \frac{1}{n!} < 1 + 1 + \tfrac{1}{2} + \tfrac{1}{2}^2 + \cdots + \frac{1}{2^n}$$

$$= 1 + 2(1 - 2^{-n}) < 3.$$

7. Let $x_1 + x_2 + \cdots + x_n = s_n$; then s_n is positive and increasing, thus either $s_n \to s > 0$ or $s_n \to +\infty$ as $n \to \infty$.
In first case $x_n = s_n - s_{n-1} \to 0$ as $n \to \infty$; in the second case

$$\frac{x_n}{s_n} \leqslant \frac{K}{s_n} \to 0 \text{ as } n \to \infty.$$

9. $u_n = (3u_{n-1})^{1/2} = 3^{1/2}(3u_{n-2})^{1/4} = 3^{1/2+1/4}(3u_{n-3})^{1/8}$
$= \cdots = 3^{1/2+1/2^2+\cdots+1/2^n}$.

10. $u_{n+1} - u_n = \dfrac{(2-u_n)(3+u_n)}{7+u_n}$, $n = 1, 2, \ldots$, then $x_{n+1} \gtrless x_n$ according as $2 \gtrless x_n$. Also $2 - u_n = \dfrac{4(2-u_{n-1})}{7+u_{n-1}}$ and so $2 \gtrless u_n$ according as $2 \gtrless c$. Hence (u_n) is monotonic and bounded; hence $u_n \to l$, where l is given by $l(7+l) = 6(1+l)$. Thus $l = 2$ since $u_n > 0$. When $c = 2$, $u_n = 2$ for all n.

11. See Example 2 Section 5.5.

12. $x_{n+1} - x_n = \dfrac{x_n - x_{n-1}}{\sqrt{(a^2 - x_{n-1})} + \sqrt{(a^2 - x_n)}}$; thus $x_{n+1} \gtrless x_n$ according as $x_n \gtrless x_{n-1}$, according as $x_1 \gtrless x_0$. Now $x_1 = \dfrac{x_0}{a + \sqrt{(a^2 - x_0)}}$.

Therefore $x_1 < x_0$ and so (x_n) is decreasing. Since $x_n > 0$, $x_n \to l \geqslant 0$, where l is given by $l = a - \sqrt{(a^2 - l)}$; $l = 0$, since $x_n < a$.

Exercises 5(d)

1. (a) $-1, 1$ (b) $-1, 0$ (c) $-\infty, \infty$ (d) $0, \frac{2}{3}$ (e) $\pm\frac{1}{2}$ (f) ± 1 (g) ± 1.

2. $M(n) > M(n+1)$ if x_n is greatest term of $x_n, x_{n+1}, x_{n+2}, \ldots$; otherwise $M(n) = M(n+1)$. Thus $M(n)$ is decreasing and so $\lim\limits_{n\to\infty} M(n) = \Lambda$. Λ is the largest of the limit points of (x_n).

4. Let $\overline{\lim}\, x_n = \Lambda$, $\overline{\lim}\, y_n = \Lambda'$ (see Section 5.8) and suppose $\Lambda' > \Lambda$. Then $0 < \Lambda' - \Lambda < y_n - x_n + 2\epsilon$ for an infinity of values of $n > N$. Thus $y_n - x_n \geqslant 0$ for an infinity of values of $n > N$ which contradicts $x_n > y_n$ for all n.
For example $x_n = (-1)^n + 1/n$, $y_n = (-1)^n + (n+1)/n$, then $x_n > y_n$ for an infinity of values of n but $\lim x_n < \lim y_n$.
Let $\underline{\lim}\, y_n = \lambda$; suppose $\lambda > \Lambda$, $0 < \lambda - \Lambda < y_n - x_n + 2\epsilon$ for $n > N$. Hence $y_n - x_n \geqslant 0$ which contradicts the hypothesis.

5. Assume $\overline{\lim}\, n(x_{n+1} - x_n) = -A < 0$.
Then $n(x_{n+1} - x_n) < -A + \epsilon$, for $n > N_1$;
$x_{n+1} - x_n < \dfrac{-A}{n} + \dfrac{\epsilon}{n} < 0, \qquad n > N_2;$

at is (x_n) is decreasing eventually. Also

$$_{n+1} - x_n > -\frac{(A+\epsilon)}{n} \text{ for an infinity of values of } n$$

$$> -\epsilon \geqslant 0 \text{ for an infinity of values of } n > N_3,$$

ntradicting (x_n) decreasing.

5. Let $\underline{\lim}\, x_{n+1}/x_n = \lambda$, $\overline{\lim}\, x_{n+1}/x_n = \Lambda$, then

$$(1-\epsilon) < \frac{x_{n+1}}{x_n} < \Lambda(1+\epsilon), \qquad n \geqslant N.$$

ultiply inequalities for $N, N+1, \ldots, n-1$; then

$$\frac{x_n^{1/n}}{\lambda} > \frac{x_N^{1/n}}{\lambda^{N/n}}(1-\epsilon)^{1-N/\epsilon} > 1-2\epsilon, \qquad n > N_1.$$

milarly $\dfrac{x_n^{1/n}}{\lambda} < 1 + 2\epsilon$ for $n > N_2$ and so

$(1-2\epsilon) < x_n^{1/n} < \Lambda(1+2\epsilon), n > N = \max(N_1, N_2)$.

7. $\infty, 0; 3, 1$.

8. $x_m - \epsilon < x_n < x_m + \epsilon, n > m$.
is bounded by $\max(x_m + 1, x_1, x_2, \ldots, x_{m-1})$ and
in $(x_m - 1, x_1, \ldots, x_{m-1})$.

xercises 6(a)

1. (a) $1 - (\frac{2}{3})^n$; converges to 1.
) $s_{2n-1} = 2n, s_{2n} = -2n$; infinite oscillation.
) $1 - 1/(n+1)$; converges to 1.
) $\frac{1}{2}\left[\frac{1}{2} - \frac{1}{(n+1)(n+2)}\right]$; converges to $\frac{1}{4}$.
) $\log(n+1)$; diverges.
) $s_{2n-1} = 1 - (\frac{1}{2})^{n-1} + \dfrac{x(1-x^n)}{1-x}$, $s_{2n} = s_{2n-1} + (\frac{1}{2})^n$;
nverges to $(1-x)^{-1}$ if $|x| < 1$; diverges if $|x| \geqslant 1$.
) $1 - (n+1)^{-2}$; converges to 1.
) $-\frac{1}{2} + (-1)^{n-1}/(n+1)$; converges to $-\frac{1}{2}$.
) $(1+x)[1-(1+x)^{-n}]$; converges to $1 + x$ if $x > 0$ or $x < -2$,
nverges to 0 when $x = 0$.

(j) $(1+x^2)\,[1-(1+x^2)^{-n}]$; converges to $1+x^2$ if $x \neq 0$, converges to zero if $x=0$.

2. (a) $s_n = d_{n+1} - d_1$, which is unbounded.
(b) $s_n = d_1^{-1} - d_{n+1}^{-1}$, which converges to d_1^{-1}.

3. Since $a_n \to 0$, $a_n^{-1} \not\to 0$ as $n \to \infty$.

Exercises 6(b)

1. Use the limit form of test for all except (g).
(a) C (b) C (c) D (d) D (e) C (f) C (g) C (h) D (i) C
(j) C (k) C $0<a<1$; D $a \geqslant 1$, $u_n \not\to 0$ (l) C.

2. $u_n \to 0$, hence $0<u_n<1$, $n \geqslant N$; and the first three results follow. Also $u_n < \frac{1}{2}$, $n \geqslant N'$; then $1-u_n > \frac{1}{2}$ and $u_n/(1-u_n) < 2u_n$.

3. (a) $\dfrac{u_{n+1}}{c_{n+1}} \leqslant \dfrac{u_n}{c_n} \leqslant \dfrac{u_{n-1}}{c_{n-1}} \leqslant \cdots \leqslant \dfrac{u_N}{c_N} = k$, then $u_n \leqslant kc_n$, $n \geqslant N$.

4. (a) C $\sqrt[n]{u_n} \to \frac{1}{3}$ as $n \to \infty$ (b) D $\sqrt[n]{u_n} \to \dfrac{e}{\pi - 1} > 1$.

(c) C $a<0$, $\sqrt[n]{u_n} \to e^a$ (d) C compare with $\Sigma n^{-4/3}$ (e) C $\sqrt[n]{u_n} \to \frac{1}{4}$
(f) C series $= \Sigma\left[\dfrac{1}{(2n-1)^2} + \dfrac{1}{(2n)^4}\right]$; compare with Σn^{-2}.

5. Let s_n be nth partial sum of $\Sigma f(n)$ and σ_n the partial sum of $\Sigma 2^n f(2^n$ then if $n = 2^k$, $2\sigma_k - u_1 < s_n < \sigma_k$.

6. (a) C $\alpha > 1$, D $\alpha \leqslant 1$ (b) C $\alpha > 1$, D $\alpha \leqslant 1$.

Exercises 6(c)

1. (a) C $0<x<1$ (b) C $0<x\leqslant 1$ (c) C $x>0$ (d) C $x>1$
(e) D for all x (f) C $0<x<1$
(g) C $0<x<\frac{3}{2}$, use Gauss's test for divergence at $x=\frac{3}{2}$
(h) C $0<x\leqslant 1$ (i) C $0<x<2$ (j) C $0<x<4$.

2. If $l>0$, choose k such that $0<k<l$; then $\phi(n)>k$ for $n \geqslant N$ and $d_n u_n - d_{n+1}u_{n+1} \geqslant k u_{n+1}$. Summing for $N, N+1, \ldots, n-1$ yields

$$k(s_n - s_N) \leqslant d_N u_N - d_n u_n < d_N u_N, \text{ where } s_n = \sum_{n=1}^{n} u_r.$$

ence s_n is bounded and Σu_n is convergent (Section 6.4).
$l < 0$, $\phi(n) < 0$ for $n \geqslant N'$ – use Exercise 6(b) No. 3.

4. Take $d_n = 1$. 5. Let $u_n = a_n - a_{n-1}$, and use Gauss's test.
8. (a) For example $u_n = (n \log n)^{-1}$ (b) C by comparison with Σn^{-2}.

Exercises 7(a)

1. (a) C (b) D ($u_n \nrightarrow 0$) (c) D ($u_n \nrightarrow 0$) (d) C (e) abs C (f) C
g) C (h) C (i) abs C (j) abs C (k) D (difference between convergent nd divergent series) (l) C $\left(u_n = (-1)^n \dfrac{[n+2-(-1)^n]}{(n+2)^2-1}\right)$.

2. (a) abs C $|u_n| \leqslant 2^{-n}$ (b) abs C $|u_n| \leqslant 3^{-n}$
c) D $S_{2n} = 1 + 3^{-3/2} + 5^{-3/2} + \cdots + (2n-1)^{-3/2}$
$-2^{-1/2}(1 + 2^{-1/2} + 3^{-1/2} + \cdots + n^{-1/2})$, and the second series is a partial sum of a divergent series.

3. $S = 13/24$ (difference between two G.P.s.)

$|S - S_{2n}| = \frac{4}{3}u_{2n+1} - \frac{1}{8}u_{2n} > u_{2n+1}$

$|S - S_{2n+1}| = \frac{9}{8}u_{2n+2} - \frac{1}{3}u_{2n+3} > u_{2n+2}$.

4. Let $p_n = \frac{1}{2}(|u_n| + u_n)$, $q_n = \frac{1}{2}(|u_n| - u_n)$; then $u_n = p_n - q_n$ and $|u_n| = p_n + q_n$.

6. First series converges, by the alternating series test, second series diverges (cf. Exercise 3, Section 7.1).

7. (a) abs C $|x| < 1$ (b) abs C $|x| < 1$, C $x = 1$
(c) abs C $|x| \leqslant 1$ (d) abs C $|x| < 1$
(e) abs C $|x| \leqslant 1$ (f) abs C all x
(g) D ($u_n \nrightarrow 0$) (h) abs C $|x| < 1$, C $x = -1$
(i) C $x = 0$ only (j) abs C $|x| < 1$
(k) abs C $|x| < 2$, $|x| = 2$, $|u_n|$ increasing, series diverges
(l) abs C $|x| < 1$, C $x = 1$ if $\alpha > 2$, C $x = -1$ if $\alpha > 0$ (See Example 4 Section 7.1)
(m) abs C $|x| < 2$, C $x = -2$ (n) abs C $|x| < e$
(o) abs C $|x| < 1$, C $x = -1$.

Exercises 7(b)

1. (a) C $0 < x < 2\pi$ (b) C $0 \leqslant x \leqslant 2\pi$ (c) abs C $0 \leqslant x \leqslant 2\pi$.

3. (a) C for all real x (use Example 4, Section 7.1) (b) C $0 \leqslant x \leqslant 2\pi$.

4. See Section 7.3. Let $u_n = nc_{n+k} = a_{n+k}b_{n+k}$, where $b_{n+k} = n/(n+k$ Convergence follows from test in first part of question.

$$S(k) = \sum_{k+1}^{\infty} a_n b_n = S(0) - \sum_{1}^{k} a_r b_r.$$

5. For the bounds of $\Sigma \sin rx$ see Example 2, Section 7.3.

6. (a) abs C $|z| < 1$ (b) abs C $|z| < 1$, C on $|z| = 1$ except at $z = 1$. (c) abs C $|z| \leqslant 1$ (d) abs C $|z| < 1$, C on $|z| = 1$ except $z = 1$. (e) abs C $|z| < 1$, C on $|z| = 1$ except $z = 1$.

7. The convergence of $\Sigma z_n = \Sigma(x_n + iy_n)$ implies both Σx_n and Σy_n ar convergent. Let $\arg z_n = \theta_n$, then $|y_n/x_n| = |\tan\theta_n| < K = \cot\delta$, when $|\theta_n| < \frac{1}{2}\pi - \delta$. Also $x_n > 0$; hence $\Sigma|y_n|$ converges by comparison. (*Note*: δ must be *fixed* positive number; $\Sigma[n^{-2} + (-1)^n n^{-1} i]$ is *not* absolutely convergent.

Exercises 7(c)

1. (a) $\sum_{n+1}^{2n} u_{2r} > n\dfrac{1}{(4n)^{1/2}} = \frac{1}{2}\sqrt{n} \to \infty$ as $n \to \infty$.

Hence $\sigma_{3n} = s_{2n} - \sum_{n+1}^{2n} u_{2r} \to -\infty$ as $n \to \infty$, since (s_{2n}) converges.

(b) See Section 7.5.

5. $\sqrt{[(r+1)(n+1-r)]} \leqslant \frac{1}{2}(n+2)$ and so $|p_n| > (n+1)\dfrac{2}{n+2}$.

Exercises 7(d)

1. (a) abs C $|x| < 3$, C $x = 3$ (use Example 4, Section 7.1)
(b) abs C $|x| < 5$
(c) abs C for all x
(d) abs C $|x-1| < 1$, i.e. $0 < x < 2$
(e) abs C $|x| > 1$
(f) abs C $|x-4| \leqslant 1$, i.e. $3 \leqslant x \leqslant 5$
(g) abs C $|x| < 1$, C $x = -1$
(h) abs C $|x| \leqslant 1$
(i) abs C $|x| < 4$
(j) abs C $|x| < 1$, C $x = -1$.

2. All abs C for $|z| < 1$; on $|z| = 1$
(a) C except for $z = 1$ (b) D (c) abs C (d) abs C (e) C except for $z = \pm i$.

xercises 8(a)

. $f^{(n)}(x) = 0, n \geqslant k+1$.

. Equate expansions to terms in $f^{(n)}$ and $f^{(n+1)}$ to give

$$^{(n)}(a+\theta h) - f^{(n)}(a) = \frac{h}{n+1} f^{(n+1)}(a+\theta' h).$$

pply the mean value theorem to L.H.S. and use the continuity of $f^{(n+1)}$, hat is $f^{(n+1)}(a+\theta' h) = f^{(n+1)}(a) + \epsilon$, where $\epsilon \to 0$ as $h \to 0$.

. $f, f^{(r)}, r = 1, 2, \ldots, n-1$ continuous in $[0, 1]$, $f^{(n)}$ exists in $(0, 1)$.

. $f^{(n)}(a+\theta_n h) = f^{(n)}(a) + \dfrac{h}{n+1} f^{(n+1)}(a) + \dfrac{h^2}{(n+1)(n+2)} f^{(n+2)}(a)$,

sing No. 3. Expand L.H.S. to give

$$_n = \frac{1}{n+1} + \frac{\frac{1}{2}h}{n+1}\left(\frac{2}{n+2} - \frac{\theta_n}{n+1}\right)\frac{f^{(n+2)}(a)}{f^{(n+1)}(a)},$$

nd use the first approximation $\theta_n \approx 1/(n+1)$.

Exercises 8(b)

2. Use induction.

3. Suppose $e = p/q$. Then

$$e = \sum_{r=0}^{q} \frac{1}{r!} + \sum_{q+1}^{\infty} \frac{1}{r!}.$$

Hence

$$q!\left\{\frac{p}{q} - \sum_{r=0}^{q} \frac{1}{r!}\right\} < \frac{1}{q}$$

which is impossible since L.H.S. must be an integer.

8. (a) 3 (b) -1 (c) $\frac{1}{4}$ (d) $-1/413$.

9. Use $2 \sin x \operatorname{sh} x = \operatorname{Im}\,(e^{(1+i)x} - e^{-(1-i)x})$.

10. By induction $g^{(n)}(x) = P(x)g(x)$, where $P(x)$ is a polynomial in $1/x$ of degree $3n$.

Exercises 8(c)

1. Use Cauchy remainder.

2. Use Lagrange remainder. Put $x = -t, 0 < t < 1$.

4. Put $x = (y-1)/(y+1)$ in the Maclaurin series for $\log[(1+x)/(1-x$

5. In second part put $x = 1/n$.

7. Cauchy's mean value theorem with $g(x) = \log x$.

9. (a) $\frac{1}{2}$ (b) $1\frac{1}{2}$ (c) $e^{-1/2\alpha^2}$ (d) $-\frac{1}{2}$.

Exercises 8(d)

1. (a) $1 - x + x^2 - \cdots + (-1)^n x^n + \cdots, |x| < 1$
(b) $1 - 2x + 3x^2 - \cdots + (-1)^n(n+1)x^n + \cdots, |x| < 1$

(c) $1 + \frac{1}{2}x - \frac{1}{8}x^2 + \cdots + (-1)^{n-1}\dfrac{1.1.3.5\cdots(2n-3)}{2.4.6.8\cdots(2n)}x^n + \cdots, |x| < 1$

(d) $1 - \frac{1}{2}x + \frac{3}{8}x^2 - \cdots + (-1)^n\dfrac{1.3.5\cdots(2n-1)}{2.4.6\cdots(2n)}x^n + \cdots, |x| < 1$.

2. (a) 8th term, $720(5/7)^7$ (b) 1st term, 1 (c) Equal 8th and 9th, $36(4/5)^7$.

3. (a) $\sqrt{3} - 1$ (b) $\frac{2}{3}(1-2x)^{3/2} - \frac{2}{3} + 2x$ (c) $\frac{1}{2}[(1-x)^{-1/2} + (1+x)^{-}$

6. (a) $-1/8$

(b) $\frac{1}{2}e$ Using Section 4.10, the limit is

$$\lim_{x\to\infty}\left(1+\frac{1}{x}\right)^x \cdot \lim_{x\to\infty}\left(x^2\log\left(1+\frac{1}{x}\right) - \frac{x}{1+1/x}\right).$$

Now use series in second limit.

Exercises 8(e)

2. (a) $1 + x + \frac{3}{2}x^2 + \frac{3}{2}x^3 + \frac{37}{24}x^4 + \cdots$ (b) $1 + \frac{1}{2}x^2 + \frac{3}{8}x^4 + \cdots$.

3. $-\frac{1}{5}$

5. $-1 + \frac{1}{8}\pi^2x^2 - \frac{1}{16}\pi^2x^3 + \dfrac{\pi^2(15-\pi^2)}{384}x^4 + \cdots$.

7. (a) $-1/36$ (b) 2 (c) $8/3$ (d) $-1/6$ (e) $1/8$ (f) $1-\frac{1}{2}\pi$
(g) $-1/6$ (h) $e^{-1/6}$
(i) -1; $(\sin x)^x = e^{x\log\sin x} = x^x(1-\frac{1}{6}x^3+\cdots)$,
$x^{\sin x} = x^x(1-\frac{1}{6}x^3\log x+\cdots)$
$(\sin x)^{\sin x} = x^x[1-\frac{1}{6}x^3(1+\log x)+\cdots]$.

8. $\sum_{n=1}^{\infty}(-1)^{n-1}(1+\frac{1}{2}+\frac{1}{3}+\cdots+1/n)x^n$, $|x|<1$.

Exercises 9(a)

1. In each case except (d), take a net with equal subdivisions (see Example 2, Section 9.3) in (d) take $x_r = ah^r$ (see Example 3, Section 9.3):
(a) $s = S = k(b-a)$

(b) $s = h^3\sum_{r=0}^{n-1} r^2$; $S = h^3\sum_{r=1}^{n} r^2 = \frac{1}{6}h^3 n(n+1)(2n+1)$

(c) $s = h\sum_{1}^{n}(1+r^2h^2)^{-1}$; $S = h\sum_{0}^{n-1}(1+r^2h^2)^{-1}$

(d) $s = n[1-{}^n\sqrt{(a/b)}]$; $S = n[\sqrt{(b/a)}-1]$
(e) $x = 1, 2, 3$ must be points of the net; $s = S = 1+2+3 = 6$

(f) $s = he^a\sum_{0}^{n-1} e^{rh} = (e^a-e^b)h/(1-e^h)$; $S = he^h(e^a-e^b)/(1-e^h)$

(g) $s = h\sum_{0}^{n-1}\sin rh = \operatorname{Im}\dfrac{(1-i)h}{1-e^{ih}} = \frac{1}{2}h(\cos\frac{1}{2}h-\sin\frac{1}{2}h)\operatorname{cosec}\frac{1}{2}h$;

$S = \frac{1}{2}h(\cos\frac{1}{2}h+\sin\frac{1}{2}h)\operatorname{cosec}\frac{1}{2}h$.

2. (a) $k(b-a)$ (b) $\lim_{n\to\infty}\frac{1}{6}a^3\left(1+\dfrac{1}{n}\right)\left(2+\dfrac{1}{n}\right) = \frac{1}{3}a^3$
(d) $\log(b/a)$ (see Exercise 4(e) No. 17) (f) $e^b - e^a$ (see Section 4.10).

3. (a) $\int_0^1 [x(1-x)]^{1/2}\,dx = \frac{1}{4}\pi$ (b) $\int_0^1 \dfrac{dx}{1+x} = \log 2$

(c) $\int_0^1 \dfrac{dx}{(1+x)^3} = \frac{3}{8}$.

4. (a) $S = n^3\sum_{0}^{n-1}(n^2+r^2)^{-2} \to \int_0^1 (1+x^2)^{-2}\,dx = (\pi+2)/8$

(b) $n^3(n+1)^3 \sum_{r=1}^{n} \frac{2r}{[n^2(n+1)^2+r^2(r-1)^2]^2}$.

Put $r=m-1$ to give half upper sum (b).

Exercises 9(b)

5. Let $I=\sup \int_a^b \alpha(x)\,dx$ for all $\alpha(x) \leqslant f(x)$,

and $J=\inf \int_a^b \beta(x)\,dx$ for all $\beta(x) \geqslant f(x)$.

Then $I \leqslant J$, and since $\int_a^b (\alpha-\beta)\,dx<\epsilon$, then $I=J$,

Choose α, β so that $I-\frac{1}{2}\epsilon<\int_a^b \alpha(x)\,dx$ and $\int_a^b \beta(x)\,dx<J+\frac{1}{2}\epsilon$.

Then for net of norm $<\delta$,

$I-\epsilon<\sum_1^n \alpha(x)(x_r-x_{r-1})<\sum_1^n f(\xi_r)(x_r-x_{r-1})<\sum_1^n \beta(x)(x_r-x_{r-1})$

$<J-\epsilon$.

Exercises 9(c)

5. Use properties 4 and 5, Section 9.6.

7. Let $G(x)=\int_a^x g(t)\,dt$. Then $\int_a^b fg\,dx=f(b)G(b)-\int_a^b f'G\,dx$,

integrating by parts. Use mean value theorem for integrals (property 5, Section 9.6).

8. As in No. 7; use $m \leqslant G \leqslant M$ and $f'<0$.
For Bonnet's form use the continuity of $G(x)$.

10. $e^t<\sqrt{(e^{2t}+e^{-t})}<\sqrt{\{e^t(e^t+\frac{1}{2})\}}$; use Properties 3 and 6, Section 9.6

11. In Schwartz's inequality take $f(\theta)=\sqrt{(1-k^2\sin^2\theta)}, g(\theta)=1$.

13. If $f(\xi)=k>0$ for a point ξ in (a, b), then since f is continuous $\exists\, \delta>0$ such that $f(x)>\frac{1}{2}k$ for $\xi-\delta<x<\xi+\delta$.

Hence $\int_a^b f(x)\,dx > \int_a^b \frac{1}{2}k\,dx = k\delta > 0$, contradicting hypothesis.

5. Use property 5, Section 9.6.

Exercises 9(d)

1. See Section 9.5; $\left|\sigma - \int_a^b f(x)\,dx\right| < \epsilon$, $\sigma = \sum_{\pi} f(\xi_r)(x_r - x_{r-1})$.

$\psi(b) - \psi(a) = \Sigma\,[\psi(x_r) - \psi(x_{r-1})] = \Sigma\psi'(\xi_r)(x_r - x_{r-1}) = \sigma$.

2. $f(b) - f(a) = \Sigma\,[f(x_r) - f(x_{r-1})] = \Sigma f'(\xi_r)(x_r - x_{r-1})$,
using the mean value theorem.

3. By No. 2 $\int_\epsilon^1 f'(x)\,dx = f(1) - f(\epsilon)$, $0 < \epsilon < 1$. Let $\epsilon \to 0$.

Exercises 9(e)

1. (a) $\frac{1}{2}$ (b) divergent (c) 3 (d) divergent (e) divergent (f) $\frac{1}{2}\pi$
(g) divergent (h) convergent to $1/a$ if $a > 0$; divergent $a \leqslant 0$
(i) $2\,[\tan^{-1} e^x]_{-\infty}^{\infty} = \pi$ (j) $\dfrac{1}{2\sqrt{2}}\tan^{-1} 2\sqrt{2}$ (k) π.

2. Cf. Section 5.5, Corollary. 3. Cf. Exercise 7(a) No. 5.

4. $\int_{-X}^{X} f(x)\,dx = 0$ since f is odd. 5. Cf. Section 9.9.

6. Use second mean value theorem, Exercise 9(c) No. 7.

7. (a) C (b) abs C (c) abs C (d) D (e) C (f) C (g) D (h) D
(i) C ($\alpha = \frac{1}{2}$) (j) C $\alpha > 0, \beta > 0$ (k) C (put $x = 1/u$ in Example 4, Section 9.8) (l) abs C $\alpha > 1$, C $0 < \alpha \leqslant 1$ (see Example 4, Section 9.8)
(m) C (put $x^2 = u$ in (l)).

8. Integral converges to $\sum_1^\infty (n+1)^{-2} = \frac{1}{6}\pi^2 - 1$.

9. $$\left.\begin{aligned} f(x) &= n^4(x - n + n^{-3}),\ n - n^{-3} \leqslant x \leqslant n \\ &= n^4(n + n^{-3} - x),\ n \leqslant x \leqslant n + n^{-3} \end{aligned}\right\}\ n = 1, 2, 3, \ldots$$
$$= 0, \text{otherwise.}$$

10. $f' \leqslant 0, \int_a^\infty |f'(x)|\,dx = \int_a^\infty f'(x)\,dx = f(a).$

Integrate $\int_a^X f(x)\cos\lambda x\,dx$ by parts.

Exercises 9(f)

1. (a) C $\alpha > 1$ (b) C (c) D (d) C (e) C $\alpha > 1$ (f) D.

4. (a) $\log\frac{3}{2}$ (b) $\log 2$ (c) $\frac{3}{2}\log 2$.

Exercises 10(a)

1. (a) Repeated limits both zero; double limit does not exist (try $y = mx$
(b) Repeated and double limits all zero (see No. 5 below)
(c) $\lim_y \lim_x f = 0$; other repeated limit and so double limit do not exist.
(d) Repeated and double limits all zero (see also No. 3)
(e) $\lim_x \lim_y e^{y\log|x|} = 1$; other repeated limit and so double limit do not exist
(f) $\cot^{-1}[(x^2+y^2)^{-1}] = \tan^{-1}(x^2+y^2)$; repeated and double limits all zero.

2. (a) $\lim_x \lim_y f = 1$, $\lim_y \lim_x f = 0$; double limit does not exist
(b) repeated limits both zero; double limit does not exist
(c) repeated and double limits all zero.

3. (a) $\overline{\lim_{x\to 0}}\, f = |y|$, $\underline{\lim_{x\to 0}}\, f = -|y|$; repeated limits both zero
(b) $\overline{\lim_{x\to 0}}\, f = |y|$, $\underline{\lim_{x\to 0}}\, f = -|y|$; repeated limits both zero
(c) As (b).

4. (a) continuous (b) discontinuous (put $y = mx$)
(c) discontinuous (See Exercise 3 Section 10.2)
(d) continuous (See No. 1(b)) (e) discontinuous
(f) discontinuous (put $y = x + x^k$, $k = 1, 2$) (g) continuous
(h) continuous (i) continuous.

5. $|f(x,y) - f(0,0)| = |f(r\cos\theta, r\sin\theta) - f(0,0)| < \epsilon$,
for $0 < r < \eta$ uniformly in θ.

6. Use No. 5.

7. Take $\epsilon = \frac{1}{2}f(a, b)$ in the condition for continuity.

Exercises 10(b)

1. (a) $\exp(x-2y); -2\exp(x-2y)$ (b) $(x-y^2)^{-1}; -2y(x-y^2)^{-1}$
(c) $\frac{1}{y}\sec^2(x/y); -\frac{x}{y^2}\sec^2(x/y)$ (d) $xy^3(1-x^2y^2)^{-3/2}; (1-x^2y^2)^{-3/2}$
(e) $2x \operatorname{sh} 2(x^2-y^2); -2y \operatorname{sh} 2(x^2-y^2)$
(f) $e^x[(x+1)\cos y - y\sin y]; -e^x[y\sin y + (x+1)\sin y]$
(g) $2y(x+y)^{-2}; -2x(x+y)^{-2}$ (h) $\frac{1}{2}x^{-1}; -\frac{1}{2}y^{-1}$
(i) $e^{xy^2}(y^2\sin x^2y + 2xy\cos x^2y); e^{xy^2}(2xy\sin x^2y + x^2\cos x^2y)$.

2. See Example 3, Section 10.5.

3. (a) $f_x(0,0) = 0 = f_y(0,0)$ (b) $f_y(0,0) = 0, f_x(0,0)$ does not exist
(c) $f_x(0,0) = 0 = f_y(0,0)$.
f_x, f_y discontinuous at $(0,0)$ in each case.

4. Let $h = \rho\cos\theta, k = \rho\sin\theta$; then the condition for differentiability is

$$\rho^{2k-1}|\sin\theta\cos\theta| = A\cos\theta + B\sin\theta + \epsilon,$$

where $\epsilon \to 0$ as $\rho \to 0$.
Let $\rho \to 0$, then if $k > \frac{1}{2}, A\cos\theta + B\sin\theta = 0$ is true for all θ if $A = 0 = B$.

5. $f_x(0,0) = 0 = f_y(0,0); f_{xx}(0,0) = f_{yy}(0,0) = f_{yx}(0,0) = 0;$
$f_{xy}(0,0)$ does not exist.

6. See Section 10.6; when $\rho \to 0$ we obtain $A\cos\theta + B\sin\theta = 0$.

7. $f(a+h, b+k) - f(a, b) = [f(a+h, b+k) - f(a, b+k)]$
$+ [f(a, b+k) - f(a, b)]$ and use the mean value theorem.

8. As in No. 7; since f_x is continuous at (a, b),
$f_x(a+\theta h, b+k) = f_x(a, b) + \epsilon_1$, where $\epsilon_1 \to 0$ as $|h| + |k| \to 0$.

9. $f_{xx}, f_{yy}, f_{xy} (= f_{yx})$ given in each case:
(a) $\exp(x-2y), 4\exp(x-2y), -2\exp(x-2y)$
(b) $-(x-y^2)^{-2}, -2(x+y^2)(x-y^2)^{-2}, 2y(x-y^2)^{-2}$
(c) $2y^{-2}\sec^2(x/y)\tan(x/y), 2xy^{-4}\sec^2(x/y)[x\tan(x/y) + y]$
$y^{-2}\sec^2(x/y)[2\tan(x/y) - 1]$.

10. $f_{xy}(0,0) = 1, f_{yx}(0,0) = -1$.

11. $f_{xy}(0,0) = 1, f_{yx}(0,0) = -1$.

12. See Section 10.8; apply mean value theorem to $\phi(x)$ and to $\psi(y)=f(x+h,y)-f(x,y)$.

13. $\Delta^2 f=\phi(a+h)-\phi(a)=h[f_x(a+\theta_1 h,b+k)-f_x(a+\theta_1 h,b)]$
$(0<\theta_1<1)$
$=h[f_{xx}(a,b).\theta_1 h+f_{yx}(a,b)k+\epsilon_1\rho-f_{xx}(a,b)\theta_1 h-\epsilon_2\rho]$
(f_x differentiable)
$=hkf_{yx}(a,b)+\epsilon\rho$, where $\epsilon_1,\epsilon_2\to 0$ and $\epsilon=h(\epsilon_1-\epsilon_2)\to 0$ as $\rho\to 0$.

Exercises 10(c)

3. $y[(y+2xt)\cos t+(2x-yt)\sin t]\exp(xy^2)$.

4. $\partial z/\partial v=0$; $z=f(x^2-y^2-2xy)$, where f is an arbitrary function.

5. $\dot{x}=\dot{r}\cos\theta-r\dot{\theta}\sin\theta$, $\dot{y}=\dot{r}\sin\theta+r\dot{\theta}\cos\theta$.

11. $\partial^2 y/\partial u\,\partial v=0$; $y=f(x-ct)+g(x+ct)$, where f, g are arbitrary functions.

12. Write $df/du=p$, then integrate to give

$$f(u)=A\int\exp(-\tfrac{1}{4}u^2/k)\,du+B,$$

where A, B are constants.

Exercises 10(d)

1. (a) min $(-1,-2)$ (b) saddle point $(0,0)$
(c) max $(-1,-1)$; saddle point $(0,0)$
(d) min $(\pm\frac{3}{2},\frac{3}{2})$; saddle point $(0,0)$, f changes sign along $x=0$
(e) max $(0,0)$; min $(\sqrt{5},-\sqrt{5}),(-\sqrt{5},\sqrt{5})$; saddle points $(1,1),(-1,-1)$
(f) min $(0,0)$; saddle points $(\operatorname{sh}\alpha,\operatorname{sh}\alpha),(-\operatorname{sh}\alpha,-\operatorname{sh}\alpha)$
(g) min $(0,0)$; max $(\frac{1}{2}\pi,0)$; saddle points $(0,\frac{1}{2}\pi),(\alpha,\alpha),(\pi-\alpha,\pi-\alpha)$ where $\alpha=\tan^{-1}\sqrt{2}$.

3. $\lambda^2<4$. 4. min $(1,2,2)$

7. $(3b,3b,-5b),(3b,-5b,3b),(-5b,3b,3b)$ where $b=a/\sqrt[3]{19}$.

ndex